AP® Teacher's Resource Gu

The Unity and Diversity of Life

FIFTEENTH EDITION

Starr

Taggart

Evers

Starr

Prepared by

Jennifer Brakeman, PhD
Head-Royce School, Oakland, CA

Kate Silber
Highland Park High School

Pamela Fedas
Weslyan School

Lesley Wade
Wiregrass Ranch High School

Julie Whitis
Simon Kenton High School

Australia • Brazil • Mexico • Singapore • United Kingdom • United States

WCN: 01-100-101

National Geographic Learning/Cengage Learning is pleased to offer our college-level materials to high schools for Advanced Placement®, honors, and electives courses. To contact your National Geographic Learning representative, please call us toll-free at **1-888-915-3276** or visit us at **http://ngl.cengage.com**.

For permission to use material from this text or product, submit all requests online at **www.cengage.com/permissions**. Further permissions questions can be emailed to **permissionrequest@cengage.com**.

ISBN: 978-1-337-40860-8

Cengage Learning
20 Channel Center Street
Boston, MA 02210
USA

Cengage Learning is a leading provider of customized learning solutions with office locations around the globe, including Singapore, the United Kingdom, Australia, Mexico, Brazil, and Japan. Locate your local office at: **www.cengage.com/global.**

Cengage Learning products are represented in Canada by Nelson Education, Ltd.

To learn more about Cengage Learning Solutions, visit **www.cengage.com**.

Printed in the United States of America
Print Number: 01 Print Year: 2018

Table of Contents

Introduction to Teaching AP® Biology

Teaching AP® Biology can be both rewarding and challenging. Students must understand that the curriculum for this course is determined by universities, and students are expected to spend more time on the material outside of class than they might for a regular high school class. The workload may be more rigorous than expected; the burden is now on the learner to find the connections. It is the intent of this guide to provide you, as the educator, with ample resources and options to direct the course. Much of the material included in this course utilizes inquiry-based techniques. Inquiry learning returns the sense of discovery back to the classroom, directing students away from simply having a list of facts and correct answers. Concepts are presented in a broader context. Biology is not a discipline where facts are simply memorized; rather concepts are studied and one applies the ideas to a variety of situations. "What if" questions are asked and experiential knowledge is used to construct new information. Interrelationships between concepts can be observed, analyzed, interpreted, and lead to further investigations.

This guide supports the content required by the College Board and is aligned with Starr/Taggart's, *Biology: The Unity and Diversity of Life*, 15th edition.

Each unit within this guide contains the following sections and format:

- Chapter correlations
- AP® Big Idea correlations
- Brief chapter summaries as they relate to the AP® Biology curriculum
- Practice multiple-choice and grid-in questions with answers
- Practice free-response questions both multiple part and single part with answer rubrics
- Review activities, a discussion of common problems, and additional resources
- Practice AP® Essays
- A section dedicated to activities and projects to promote learning after the exam
- Some or all of the following chapter-specific information
    - Objectives
    - AP® Enduring Understandings and Enduring Knowledge
    - Warm Up Questions
    - Lesson Openers and Closers
    - Suggestions for Presenting the Material
    - Classroom and Laboratory Enrichment (which includes the suggested AP® labs)
    - Classroom Discussion and Activities Ideas
    - Homework Extension Questions
    - Possible Responses to *Critical Thinking* Questions
    - Possible Responses to *Data Analysis Activities* Questions

---

AP® is a trademark registered by the College Board, which is not affiliated with, and does not endorse, this product.

School calendars vary, but the timeline provided in lesson outlines will help with course planning in your particular school setting. Whether a veteran or rookie teacher, having the foundation of a lesson is always helpful; working out a daily plan is then left to your own style. This guide will assist in streamlining the learning to fit what is expected by the College Board. Inquiry-based learning can easily be incorporated into the material, both in the lecture and the lab. Practice questions in the format of the new AP® Exam are included in this text, along with multiple-choice, grid-in, and multiple-part and single-part free-response questions and answers. These are provided so that students may practice writing answers and getting used to the testing format for the exam. The Data Analysis Exercises at the end of each chapter in the textbook can also be used along with these resources as great practice tools in the classroom setting to build scientific thinking skills and prepare students for the time constraints of the exam. This will build confidence and increase their ability to perform under pressure on the actual test.

The review activities provide extensions to the material in a way that will encourage students to embody the concepts so that comprehension and understanding are at the forefront of the learning. Once the AP® Exam has been taken, students should still have the opportunity to learn in AP® Biology. "What to do after the Exam?" will provide opportunities after the course work is done and there are still school days to fill with instruction. Allowing the student some flexibility in choosing the area that interests them most is now something that is reasonable. Independent study may also be introduced at this time, so students might gain insight in concepts not covered in the curriculum of the course. This is just another way to direct the student toward learning processes utilized at the college level.

Suggested Guide for Using the AP® Teacher's Resource Guide for AP® Biology

The unit numbers used in this guide correspond with the textbook unit numbers for ease of use. There is no Unit 4 in this teacher resource guide (TRG) because Unit 4 in the textbook does not correlate with content that needs to be taught in the AP® Biology curriculum, other than in the context of some illustrative examples.

TRG Unit	TRG Unit Title	Book Unit	Book Chapters	AP® Biology Investigations
1A	Principles of Cellular Life: Biochemistry	1	1–4	4 (proc 1)
1B	Principles of Cellular Life: Metabolism		5–7	4 (proc 2–3), 5, 6, 13
2A	Genetics: Genes and Gene Expression and Biotechnology	2	8–10, 15	8, 9
2B	Genetics: Cell Reproduction and Mendelian Genetics		11–15	7
3	Evolution	3	16–19	1, 2, 3
5	Focus on Plants	5	22, 28, 30	1, 5, 11
6	Organisms	6	32–34, 37, 41–42	
7	Principles of Ecology	7	43–48	10, 12

You can find the AP® investigations suggested by the College Board at this link: https://apcentral.collegeboard.org/courses/ap-biology/course/lab-manual-resource-center

AP® Investigation Correlation Chart Organized by Investigation

<table>
<tr><th>AP® Big Idea</th><th>Investigation #</th><th>AP® Investigation Title</th><th>IRM Unit Correlation(s)</th><th>Textbook Unit Correlation(s)</th><th>Textbook Chapter Correlation(s)</th></tr>
<tr><td rowspan="3">1: Evolution</td><td>1</td><td>Artificial selection</td><td>3, 5</td><td>3</td><td>17</td></tr>
<tr><td>2</td><td>Mathematical modeling: Hardy-Weinberg</td><td>3</td><td>3</td><td>17</td></tr>
<tr><td>3</td><td>Comparing DNA sequences to understand evolutionary relationships with BLAST</td><td>2, 3</td><td>2, 3</td><td>15, 17</td></tr>
<tr><td rowspan="4">2. Cellular Processes: Energy and Communication</td><td rowspan="2">4</td><td>Diffusion and osmosis, Procedure 1: Surface area and cell size</td><td>1A</td><td>1</td><td>4</td></tr>
<tr><td>Diffusion and osmosis, Procedures 2 and 3: Modeling and observing in living cells</td><td>1B</td><td>1</td><td>5.7–5.10</td></tr>
<tr><td>5</td><td>Photosynthesis</td><td>1B, 5</td><td>1</td><td>6, 7, 46</td></tr>
<tr><td>6</td><td>Cellular respiration</td><td>1B</td><td>1</td><td>7</td></tr>
<tr><td rowspan="3">3: Genetics and Information Transfer</td><td>7</td><td>Cell division: Mitosis and meiosis</td><td>2B</td><td>2</td><td>11, 12</td></tr>
<tr><td>8</td><td>Biotechnology: Bacterial transformation</td><td>2B</td><td>2</td><td>15</td></tr>
<tr><td>9</td><td>Biotechnology: Restriction enzyme analysis of DNA</td><td>2B</td><td>2</td><td>15</td></tr>
<tr><td rowspan="4">4. Interactions</td><td>10</td><td>Energy dynamics</td><td>7</td><td>7</td><td>46</td></tr>
<tr><td>11</td><td>Transpiration</td><td>5</td><td>5</td><td>28</td></tr>
<tr><td>12</td><td>Fruit fly behavior</td><td>6</td><td>6</td><td>33, 43</td></tr>
<tr><td>13</td><td>Enzyme activity</td><td>1B</td><td>1</td><td>5.1–5.6</td></tr>
</table>

Topic Correlation Chart

Big Idea 1: The process of evolution drives the diversity and unity of life.	
Enduring Understanding 1.A: Change in the genetic makeup of a population over time is evolution.	
Essential Knowledge (EK)	**Chapter/Section**
EK1.A.1: Natural selection is a major mechanism of evolution.	**1.4** How Things Differ **16.1** Early Beliefs, Confounding Discoveries; **16.3** A Flurry of New Theories; **16.4** Darwin, Wallace, and Natural Selection **17.1** Superbug Farms; **17.2** Individuals Don't Evolve, Populations Do; **17.4** Patterns of Natural Selection; **17.5** Directional Selection; **17.6** Stabilizing and Disruptive Selection; **17.7** Fostering Diversity; **17.11** Other Speciation Models **22.2** Plant Ancestry and Diversity; **22.3** Evolutionary Trends Among Plants; **22.9** Angiosperm Diversity and Importance **23.2** Fungal Traits and Classification; **23.7** Ecological Roles of Fungi **24.1** Medicines from the Sea; **24.2** Animal Traits and Body Plans; **24.3** Animal Origins and Adaptive Radiation **25.1** Very Early Birds; **25.2** Chordate Traits and Evolutionary Trends; **25.4** Evolution of Jawed Fishes; **25.7** Amniote Evolution **26** All sections **31.2** Organization of Human Bodies **41.2** Modes of Animal Reproduction **43.2** Behavioral Genetics; **43.6** Communication Signals; **43.7** Mates, Offspring, and Reproductive Success; **43.9** Why Sacrifice Yourself? **45.6** Evolutionary Arms Races; **45.9** Species Introduction, Loss, and Other Disturbances
EK1.A.2: Natural selection acts on phenotypic variations in populations.	**1.4** How Living Things Differ **12.1** Why Sex?; **12.4** How Meiosis Introduces Variations in Traits **16.1** Early Beliefs, Confounding Discoveries; **16.3** A Flurry of New Theories; **16.4** Darwin, Wallace, and Natural Selection; **16.5** Fossils: Evidence of Ancient Life **17.4** Patterns of Natural Selection; **17.5** Directional Selection; **17.6** Stabilizing and Disruptive Selection; **17.7** Fostering Diversity; **17.11** Other Speciation Models **22.2** Plant Ancestry and Diversity; **22.3** Evolutionary Trends Among Plants **23.2** Fungal Traits and Classification **24.2** Animal Traits and Body Plans; **24.3** Animal Origins and Adaptive Radiation **25.2** Chordate Traits and Evolutionary Trends; **25.4** Evolution of Jawed Fishes; **25.7** Amniote Evolution **26** All sections **31.2** Organization of Animal Bodies **41.2** Modes of Animal Reproduction **43.2** Behavioral Genetics; **43.6** Communication Signals; **43.7** Mates, Offspring, and Reproductive Success; **43.9** Why Sacrifice Yourself? **45.6** Evolutionary Arms Races; **45.9** Species Introduction, Loss, and Other Disturbances
EK 1.A.3: Evolutionary change is also driven by random processes.	**14.5** Heritable Changes in Chromosome Structure **16.4** Darwin, Wallace, and Natural Selection **17.8** Genetic Drift and Gene Flow; **17.9** Reproductive Isolation; **17.10** Allopatric Speciation

EK 1.A.4: Biological evolution is supported by scientific evidence from many disciplines, including mathematics.	**16.5** Fossils: Evidence of Ancient Life; **16.6** Filling in Pieces of the Puzzle; **16.7** Drifting Continents, Changing Seas **17** All sections **18.3** Comparing Form and Function; **18.4** Comparing Biochemistry; **18.5** Comparing Patterns of Animal Development **19** All sections **25.1** Very Early Birds **26** All sections

Enduring Understanding 1.B: Organisms are linked by lines of descent from common ancestry.	
EK 1.B.1: Organisms share many conserved core processes and features that evolved and are widely distributed among organisms today.	**4.2** What Is a Cell? **14.5** Heritable Changes in Chromosome Structure **19** All sections **20.5** Shared Traits of Prokaryotes; **20.6** Factors in the Success of Prokaryotes; **20.7** A Sample of Bacterial Diversity; **20.8** Archaea **21–26, 30** All sections **31.2** Organization of Human Bodies; **31.8** Human Integumentary System **32.2** Evolution of Nervous Systems; **32.10** The Vertebrate Brain **33** All sections **34.2** The Vertebrate Endocrine System; **34.13** Invertebrate Hormones **35.2** Animal Movement; **35.3** The Vertebrate Endoskeleton **36.2** Circulatory Systems **37.2** Integrated Responses to Threats **38.2** The Nature of Respiration; **38.3** Invertebrate Respiration; **38.4** Vertebrate Respiration **39.2** Animal Digestive Systems **40.2** Regulating Fluid Volume and Composition **41.2** Modes of Animal Reproduction **42.2** Stages of Development; **42.3** From Zygote to Gastrula; **42.4** How Specialized Tissues and Organs Form; **42.5** An Evolutionary View of Development **43** All sections **45.6** Evolutionary Arms Races
EK 1.B.2: Phylogenetic trees and cladograms are graphical representations (models) of evolutionary history that can be tested.	**1.5** Organizing Information About Species **18** All sections **21.2** A Collection of Lineages **22.2** Plant Ancestry and Diversity **23.2** Fungal Traits and Classification **24.2** Animal Traits and Body Plans **25.2** Chordate Trends and Evolutionary Trends **26.2** Primates: Our Order

Enduring Understanding 1.C: Life continues to evolve within a changing environment.	
EK 1.C.1: Speciation and extinction have occurred throughout the Earth's history.	**16** All sections **17.10** Allopatric Speciation; **17.11** Other Speciation Models; **17.12** Macroevolution **18.1** Bye Bye Birdie **22–23** All sections **24.3** Animal Origins and Adaptive Radiation **25.2** Chordate Traits and Evolutionary Trends; **25.4** Evolution of Jawed Fishes; **25.7** Amniote Evolution **26** All sections **44.6** Effects of Predation on Life History **48** All sections
EK 1.C.2: Speciation may occur when two populations become reproductively isolated from each other.	**17.9** Reproductive Isolation **26.6** Recent Human Lineages
EK 1.C.3: Populations of organisms continue to evolve.	**17.1** Superbug Farms; **17.2** Individuals Don't Evolve, Populations Do **20.1** Evolution of a Disease; **20.4** Viruses as Human Pathogens **41.2** Modes of Animal Reproduction **44.5** Life History Patterns; **44.6** Effects of Predation on Life History **45.6** Evolutionary Arms Races; **45.9** Species Introduction, Loss, and Other Disturbances **46.8** Greenhouse Gases and Climate Change **47.5** Biomes; **47.6** Deserts; **47.7** Grasslands; **47.8** Dry Shrublands and Woodlands; **47.9** Broadleaf Forests; **47.10** Coniferous Forests; **47.11** Tundra; **47.12** Freshwater Ecosystems; **47.13** Coastal Ecosystems; **47.14** Coral Reefs; **47.15** The Open Ocean
Enduring Understanding 1.D: The origin of living systems is explained by natural processes.	
EK 1.D.1: There are several hypotheses about the natural origin of life on Earth, each with supporting scientific evidence.	**16, 19** All sections **22.6** History of the Vascular Plants **24.2** Animal Traits and Body Plans; **24.3** Animal Origins and Adaptive Radiation **25.1** Very Early Birds; **25.2** Chordate Traits and Evolutionary Trends; **25.4** Evolution of Jawed Fishes; **25.7** Amniote Evolution
EK 1.D.2: Scientific evidence from many different disciplines supports models of the origin of life.	**16, 19** All sections

Big Idea 2: Biological systems utilize free energy and molecular building blocks to grow, reproduce, and maintain dynamic homeostasis.	
Enduring Understanding 2.A: Growth, reproduction, and maintenance of the organization of living systems require free energy and matter.	
EK 2.A.1: All living systems require constant input of free energy.	**1.3** How Living Things Are Alike **6–7** All sections **20.5** Shared Traits of Prokaryotes; **20.6** Factors in the Success of Prokaryotes; **20.7** A Sample of Bacterial Diversity; **20.8** Archaea **21** All sections **22.4** Bryophytes; **22.5** Seedless Vascular Plants; **22.7** Gymnosperms; **22.8** Angiosperms **23.3** Flagellated Fungi; **23.4** Zygote Fungi and Related Groups; **23.5** Sac Fungi—Ascomycetes; **23.6** Club Fungi—Basidiomycetes **4.4** Sponges; **24.5** Cnidarians—Predators with Stinging Cells; **24.6** Flatworms—Simple Organ Systems; **24.7** Annelids—Segmented Worms; **24.8** Mollusks—Animals with a Mantle; **24.9** Rotifers and Tardigrades—Tiny and Tough; **24.10** Roundworms—Unsegmented Worms that Molt; **24.11** Arthropods—Molting Animals with Jointed Legs; **24.12** Chelicerates—Spiders and Their Relatives; **24.13** Myriapods; **24.14** Crustaceans; **24.15** Insects—Diverse and Abundant; **24.16** The Spiny-Skinned Echinoderms **25.3** Jawless Fishes; **25.5** Modern Jawed Fishes; **25.6** Amphibians—First Tetrapods on Land; **25.8** Nonbird Reptiles; **25.9** Birds—the Feathered Ones; **25.10** Mammals—Milk Makers **27.5** Leaves **28.2** Plant Nutrients and Availability of Soil **39** All sections **42.9** Structure and Function of the Placenta; **42.11** Milk: Nourishment and Protection **46.3** The Nature of Food Webs; **46.4** Energy Flow
EK 2.A.2: Organisms capture and store free energy for use in biological processes.	**6–7** All sections **20.5** Shared Traits of Prokaryotes; **20.6** Factors in the Success of Prokaryotes; **20.7** A Sample of Bacterial Diversity; **20.8** Archaea Chap **21** All sections **22.4** Bryophytes; **22.5** Seedless Vascular Plants; **22.7** Gymnosperms; **22.8** Angiosperms **23.3** Flagellated Fungi; **23.4** Zygote Fungi and Related Groups; **23.5** Sac Fungi—Ascomycetes; **23.6** Club Fungi—Basidiomycetes **24.4** Sponges; **24.5** Cnidarians—Predators with Stinging Cells; **24.6** Flatworms—Simple Organ Systems; **24.7** Annelids—Segmented Worms; **24.8** Mollusks—Animals with a Mantle; **24.9** Rotifers and Tardigrades—Tiny and Tough; **24.10** Roundworms—Unsegmented Worms that Molt; **24.11** Arthropods—Molting Animals with Jointed Legs; **24.12** Chelicerates—Spiders and Their Relatives; **24.13** Myriapods; **24.14** Crustaceans; **24.15** Insects—Diverse and Abundant; **24.16** The Spiny-Skinned Echinoderms **25.3** Jawless Fishes; **25.5** Modern Jawed Fishes; **25.6** Amphibians—First Tetrapods on Land; **25.8** Nonbird Reptiles; **25.9** Birds—the Feathered Ones; **25.10** Mammals—Milk Makers **39** All sections **42.11** Milk: Nourishment and Protection **44.4** Limits on Population Growth **46.3** The Nature of Food Webs; **46.4** Energy Flow
EK 2.A.3: Organisms must exchange matter with the environment to grow, reproduce, and maintain organization.	**2** All sections **4.11** Cell Surface Specializations **5.7** A Closer Look at Membranes; **5.8** Diffusion and Membranes; **5.9** Membrane Transport Mechanisms; **5.10** Membrane Trafficking **20.5** Shared Traits of Prokaryotes; **20.6** Factors in the Success of Prokaryotes; **20.7** A Sample of Bacterial Diversity; **20.8** Archaea **21** All sections **22.4** Bryophytes; **22.5** Seedless Vascular Plants; **22.7** Gymnosperms; **22.8** Angiosperms

	23.3 Flagellated Fungi; **23.4** Zygote Fungi and Related Groups; **23.5** Sac Fungi—Ascomycetes; **23.6** Club Fungi—Basidiomycetes **24.4** Sponges; **24.5** Cnidarians—Predators with Stinging Cells; **24.6** Flatworms—Simple Organ Systems; **24.7** Annelids—Segmented Worms; **24.8** Mollusks—Animals with a Mantle; **24.9** Rotifers and Tardigrades—Tiny and Tough; **24.10** Roundworms—Unsegmented Worms that Molt; **24.11** Arthropods—Molting Animals with Jointed Legs; **24.12** Chelicerates—Spiders and Their Relatives; **24.13** Myriapods; **24.14** Crustaceans; **24.15** Insects—Diverse and Abundant; **24.16** The Spiny-Skinned Echinoderms **25.3** Jawless Fishes; **25.5** Modern Jawed Fishes; **25.6** Amphibians—First Tetrapods on Land; **25.8** Nonbird Reptiles; **25.9** Birds—the Feathered Ones; **25.10** Mammals—Milk Makers **27.5** Leaves; **27.6** Roots **28.3** Root Adaptations for Nutrient Uptake **29, 38** All sections **39.2** Animal Digesting Systems; **39.3** Overview of the Human Digestive System **40.2** Regulating Fluid Volume and Composition **42** All sections **44.4** Limits on Population Growth **46** All sections **47.5** Biomes; **47.6** Deserts; **47.7** Grasslands; **47.8** Dry Shrublands and Woodlands; **47.9** Broadleaf Forests; **47.10** Coniferous Forests; **47.11** Tundra; **47.12** Freshwater Ecosystems; **47.13** Coastal Ecosystems; **47.14** Coral Reefs; **47.15** The Open Ocean
Enduring Understanding 2.B: Growth, reproduction and dynamic homeostasis require that cells create and maintain internal environments that are different from their external environments.	
EK 2.B.1: Cell membranes are selectively permeable due to their structure.	**4.2** What Is a Cell? **5.7** A Closer Look at Cell Membranes; **5.8** Diffusion and Membranes; **5.9** Membrane Transport Mechanisms; **5.10** Membrane Trafficking **20.5** Shared Traits of Prokaryotes
EK 2.B.2: Growth and dynamic homeostasis are maintained by the constant movement of molecules across membranes.	**1.3** How Living Things Are Alike **4.2** What Is a Cell?; **4.7** The Endomembrane System; **4.11** Cell Surface Specializations **5.7** A Closer Look at Cell Membranes; **5.8** Diffusion and Membranes; **5.9** Membrane Transport Mechanisms; **5.10** Membrane Trafficking **20.5** Shared Traits of Prokaryotes; **20.6** Factors in the Success of Prokaryotes **21.2** A Collection of Lineages **23.2** Fungal Traits and Classification **27.2** The Plant Body; **27.3** Plant Tissues; **27.5** Leaves; **27.6** Roots **28.3** Root Adaptations for Nutrient Uptake; **28.6** Movement of Organic Compounds in Plants **30.2** Introduction to Plant Hormones; **30.3** Auxin: The Master Growth Hormone; **30.4** Cytokinin; **30.5** Gibberellin; **30.6** Abscisic Acid; **30.7** Ethylene; **30.10** Responses to Stress **31.3** Epithelial Tissue; **31.4** Connective Tissues; **31.5** Muscle Tissues; **31.6** Nervous Tissue **32.4** Membrane Potential; **32.5** The Action Potential; **32.6** How Neurons Send Messages to Other Cells **33.3** Somatic and Visceral Sensations; **33.4** Chemical Senses; **33.7** Light Reception and Visual Processing; **33.9** Vertebrate Hearing; **33.10** Organs of Equilibrium **34** All sections **35.7** How Muscle Contracts; **35.9** Muscle Metabolism; **35.8** Nervous Control of Muscle Contraction **36.2** Circulatory Systems; **36.3** Human Cardiovascular System; **36.8** Exchanges at Capillaries; **36.11** Interactions with the Lymphatic System **37–38** All sections **39.2** Animal Digestive Systems; **39.3** Overview of the Human Digestive System; **39.36** Structure of the Small Intestine; **39.7** Digestion and Absorption in the Small Intestine; **39.8** The Large Intestine

	40.2 Regulating Fluid Volume and Composition; **40.3** The Human Urinary System; **40.4** How Urine Forms; **40.5** Fluid Homeostasis **42.2** Stages of Development; **42.3** From Zygote to Gastrula; **42.4** How Specialized Tissues and Organs Form; **42.7** Early Human Development; **42.8** Emergence of Distinctly Human Features; **42.9** Structure and Function of the Placenta; **42.10** Labor and Childbirth
EK 2.B.3: Eukaryotic cells maintain internal membranes that partition the cell into specialized regions.	**4.5** Introducing Eukaryotic Cells; **4.6** The Nucleus; **4.7** The Endomembrane System; **4.8** Mitochondria; **4.9** Chloroplasts and Other Plastids
Enduring Understanding 2.C: Organisms use feedback mechanisms to regulate growth and reproduction and to maintain dynamic homeostasis.	
EK 2.C.1: Organisms use feedback mechanisms to maintain their internal environments and respond to external environmental changes.	**1.3** Living Things Are Alike **30.2** Introduction to Plant Hormones; **30.3** Auxin: The Master Growth Hormone; **30.4** Cytokinin; **30.5** Gibberellin; **30.6** Abscisic Acid; **30.7** Ethylene; **30.10** Responses to Stress **31.9** Negative Feedback in Homeostasis **32–38** All sections **39.2** Animal Digestive Systems; **39.3** Overview of the Human Digestive System; **39.36** Structure of the Small Intestine; **39.7** Digestion and Absorption in the Small Intestine; **39.8** The Large Intestine **40** All sections **41.4** Reproductive System of Human Females; **41.5** Female Reproductive Cycles; **41.6** Reproductive System of Human Males **42.2** Stages of Development; **42.3** From Zygote to Gastrula; **42.4** How Specialized Tissues and Organs Form; **42.7** Early Human Development; **42.8** Emergence of Distinctly Human Features; **42.9** Structure and Function of the Placenta; **42.10** Labor and Childbirth; **42.11** Milk: Nourishment and Protection **43–44** All sections
EK 2.C.2: Organisms respond to changes in their external environments.	**1.3** How Living Things Are Alike **21** All sections **22.4** Bryophytes; **22.5** Seedless Vascular Plants; **22.7** Gymnosperms; **22.8** Angiosperms **23** All sections **24.4** Sponges; **24.5** Cnidarians—Predators with Stinging Cells; **24.6** Flatworms—Simple Organ Systems; **24.7** Annelids—Segmented Worms; **24.8** Mollusks—Animals with a Mantle; **24.9** Rotifers and Tardigrades—Tiny and Tough; **24.10** Roundworms—Unsegmented Worms that Molt; **24.11** Arthropods—Molting Animals with Jointed Legs; **24.12** Chelicerates—Spiders and Their Relatives; **24.13** Myriapods; **24.14** Crustaceans; **24.15** Insects—Diverse and Abundant; **24.16** The Spiny-Skinned Echinoderms **25.3** Jawless Fishes; **25.5** Modern Jawed Fishes; **25.6** Amphibians—First Tetrapods on Land; **25.8** Nonbird Reptiles; **25.9** Birds—the Feathered Ones; **25.10** Mammals—Milk Makers **27.1** Carbon Sequestration; **27.7** Primary Growth; **27.8** Secondary Growth; **27.9** Tree Rings and Old Secrets **28.5** Water-Conserving Adaptations of Stems and Leaves **30.8** Tropisms; **30.9** Sensing Recurring Environmental Change; **30.10** Responses to Stress **31—34** All sections **35.2** Animal Movement **37** All sections **38.6** How We Breathe; **38.8** Respiratory Adaptations **40, 43–45, 48** All sections

Enduring Understanding 2.D: Growth and dynamic homeostasis of a biological system are influenced by changes in the system's environment.	
EK 2.D.1: All biological systems from cells and organisms to populations, communities, and ecosystems are affected by complex biotic and abiotic interactions involving exchange of matter and free energy.	**5.2** Energy in the World of Life **20.7** A Sample of Bacterial Diversity **21** All sections **22.4** Bryophytes; **22.5** Seedless Vascular Plants; **22.7** Gymnosperms; **22.8** Angiosperms **23.3** Flagellated Fungi; **23.4** Zygote Fungi and Related Groups; **23.5** Sac Fungi—Ascomycetes; **23.6** Club Fungi—Basidiomycetes **24.4** Sponges; **24.5** Cnidarians—Predators with Stinging Cells; **24.6** Flatworms—Simple Organ Systems; **24.7** Annelids—Segmented Worms; **24.8** Mollusks—Animals with a Mantle; **24.9** Rotifers and Tardigrades—Tiny and Tough; **24.10** Roundworms—Unsegmented Worms that Molt; **24.11** Arthropods—Molting Animals with Jointed Legs; **24.12** Chelicerates—Spiders and Their Relatives; **24.13** Myriapods; **24.14** Crustaceans; **24.15** Insects—Diverse and Abundant; **24.16** The Spiny-Skinned Echinoderms **25.3** Jawless Fishes; **25.5** Modern Jawed Fishes; **25.6** Amphibians—First Tetrapods on Land; **25.8** Nonbird Reptiles; **25.9** Birds—the Feathered Ones; **25.10** Mammals—Milk Makers **27.1** Carbon Sequestration; **27.7** Primary Growth; **27.8** Secondary Growth; **27.9** Tree Rings and Old Secrets **28.3** Root Adaptations for Nutrient Uptake; **28.5** Water-Conserving Adaptations of Stems and Leaves **29** All sections **30.8** Tropisms; **30.9** Sensing Recurring Environmental Change; **30.10** Responses to Stress **39–40** All sections **42.9** Structure and Function of the Placenta; **42.10** Labor and Childbirth; **42.11** Milk: Nourishment and Protection **44–48** All sections
EK 2.D.2: Homeostatic mechanisms reflect both common ancestry and divergence due to adaptation in different environments.	**21–26, 30–31** All sections **32.2** Evolution of Nervous Systems; **32.10** The Vertebrate Brain **34.2** The Vertebrate Endocrine System; **34.13** Invertebrate Hormones **35.2** Animal Movement; **35.3** The Vertebrate Endoskeleton **36.2** Circulatory Systems; **36.5** Vertebrate Blood **37.2** Integrated Responses to Threats **38.2** The Nature of Respiration; **38.3** Invertebrate Respiration; **38.4** Vertebrate Respiration **39.2** Animal Digestive Systems **40.2** Regulating Fluid Volume and Composition **43** All sections
EK 2.D.3: Biological systems are affected by disruptions to their dynamic homeostasis.	**1.3** How Living Things Are Alike **30.8** Tropisms; **30.9** Sensing Recurring Environmental Change; **30.10** Responses to Stress **31.9** Negative Feedback in Homeostasis **33–34** All sections **35.4** Bone Structure and Function; **35.5** Bone and Joint Health; **35.8** Nervous Control of Muscle Contraction; **35.9** Muscle Metabolism **36.10** Blood and Cardiovascular Disorders **37** All sections **38.9** Respiratory Diseases and Disorders **39.1** Your Microbial Organ; **39.12** Maintaining a Healthy Weight **40.6** When Kidneys Fail; **40.7** Heat Gains and Losses; **40.8** Adaptations to Heat and Cold **41.8** Contraception and Infertility; **41.9** Sexually Transmitted Diseases **42.1** Prenatal Problems **45.1** Fighting Foreign Fire Ants **47.2** Global Air Circulation Patterns; **47.3** The Ocean, Landforms, and Climates; **47.4** The El Nino Southern Oscillation **48** All sections

EK 2.D.4: Plants and animals have a variety of chemical defenses against infections that affect dynamic homeostasis.	**30.9** Sensing Recurring Environmental Change; **30.10** Responses to Stress **31.7** Organ Systems; **31.8** Human Integumentary System **37** All sections
Enduring Understanding 2.E: Many biological processes involved in growth, reproduction, and dynamic homeostasis include temporal regulation and coordination.	
EK 2.E.1: Timing and coordination of specific events are necessary for the normal development of an organism, and these events are regulated by a variety of mechanisms.	**10.2** Between You and Eternity;**10.3** Switching Genes On and Off; **10.4** Examples of Gene Control in Eukaryotes; **10.6** Epigenetics **18.5** Comparing Patterns of Animal Development **22.4** Bryophytes; **22.5** Seedless Vascular Plants; **22.7** Gymnosperms; **22.8** Angiosperms **23.3** Flagellated Fungi; **23.4** Zygote Fungi and Related Groups; **23.5** Sac Fungi—Ascomycetes; **23.6** Club Fungi—Basidiomycetes **24.4** Sponges; **24.5** Cnidarians—Predators with Stinging Cells; **24.6** Flatworms—Simple Organ Systems; **24.7** Annelids—Segmented Worms; **24.8** Mollusks—Animals with a Mantle; **24.9** Rotifers and Tardigrades—Tiny and Tough; **24.10** Roundworms—Unsegmented Worms that Molt; **24.11** Arthropods—Molting Animals with Jointed Legs; **24.12** Chelicerates—Spiders and Their Relatives; **24.13** Myriapods; **24.14** Crustaceans; **24.15** Insects—Diverse and Abundant; **24.16** The Spiny-Skinned Echinoderms **25.3** Jawless Fishes; **25.5** Modern Jawed Fishes; **25.6** Amphibians—First Tetrapods on Land; **25.8** Nonbird Reptiles; **25.9** Birds—the Feathered Ones; **25.10** Mammals—Milk Makers **29** All sections **30.2** Introduction to Plant Hormones; **30.3** Auxin: The Master Growth Hormone; **30.4** Cytokinin; **30.5** Gibberellin; **30.6** Abscisic Acid; **30.7** Ethylene **31** All sections **32.10** The Vertebrate Brain **37** All sections **41.3** Organs of Sexual Reproduction; **41.4** Reproductive System of Human Females; **41.5** Female Reproductive Cycles; **41.6** Reproductive System of Human Males; **41.7** Bringing Gametes Together **42–43** All sections
EK 2.E.2: Timing and coordination of physiological events are regulated by multiple mechanisms.	**28–29** All sections **30.2** Introduction to Plant Hormones; **30.3** Auxin: The Master Growth Hormone; **30.4** Cytokinin; **30.5** Gibberellin; **30.6** Abscisic Acid; **30.7** Ethylene **31–34** All sections **35.4** Bone Structure and Function; **35.6** Skeleton Muscle Function; **35.7** How Muscle Contracts; **35.8** Nervous Control of Muscle Contraction; **35.9** Muscle Metabolism **36–40** All sections **41.3** Organs of Sexual Reproduction; **41.4** Reproductive System of Human Females; **41.5** Female Reproductive Cycles; **41.6** Reproductive System of Human Males; **41.7** Bringing Gametes Together **42–43** All sections
EK 2.E.3: Timing and coordination of behavior are regulated by various mechanisms and are important in natural selection.	**17.7** Fostering Diversity; **17.9** Reproductive Isolation **24–26** All sections **29.3** Flowers and Their Pollinators; **29.7** Fruits **32–33** All sections **34** All sections **39.11** What Should You Eat?; **39.12** Maintaining a Healthy Weight **40.8** Adaptations to Heat and Cold **41.5** Female Reproductive Cycles; **41.7** Bringing Gametes Together **43–45** All sections **46.2** The Nature of Ecosystems

Big Idea 3: Living systems store, retrieve, transmit, and respond to information essential to life processes.	
Enduring Understanding 3.A: Heritable information provides for continuity of life.	
EK 3.A.1: DNA, and in some cases RNA, is the primary source of heritable information.	**1.3** How Living Things Are Alike **8.2** The Discovery of DNA's Function **13.2** Mendel, Pea Plants, and Inheritance Patterns **20.2** Viruses and Viroids; **20.3** Viral Replication
EK 3.A.2: In eukaryotes, heritable information is passed to the next generation via processes that include the cell cycle and mitosis or meiosis plus fertilization.	**3.8** Nucleic Acids **8.5** DNA Replication **11–12** All sections **13.2** Mendel, Pea Plants, and Inheritance Patterns; **13.3** Mendel's Law of Segregation; **13.4** Mendel's Law of Independent Assortment **14.5** Heritable Changes in Chromosome Structure; **14.6** Heritable Changes in Chromosome Number **29.4** A New Generation Begins; **29.5** Flower Sex **41.7** Bringing Gametes Together
EK 3.A.3: The chromosomal basis of inheritance provides an understanding of the pattern of passage (transmission) of genes from parent to offspring.	**8.4** Eukaryotic Chromosomes **11.2** Multiplication by Division; **11.3** A Closer Look at Mitosis **12** All sections **13.2** Mendel, Pea Plants, and Inheritance Patterns; **13.3** Mendel's Law of Segregation; **13.4** Mendel's Law of Independent Assortment **14** All sections **41.7** Bringing Gametes Together
EK 3.A.4: The inheritance pattern of many traits cannot be explained by simple Mendelian genetics.	**13.5** Beyond Simple Dominance; **13.6** Nature and Nurture; **13.7** Complex Variation in Traits
Enduring Understanding 3.B: Expression of genetic information involves cellular and molecular mechanisms.	
EK 3.B.1: Gene regulation results in differential gene expression, leading to cell specialization.	**9–10** All sections **31.1** Stem Cells—It's All About Potential **42** All sections
EK 3.B.2: A variety of intercellular and intracellular signal transmissions mediate gene expression.	**9–10** All sections **31.9** Negative Feedback in Homeostasis **37.6** Antigen Receptors **42** All sections
Enduring Understanding 3.C: The processing of genetic information is imperfect and is a source of genetic variation.	
EK 3.C.1: Changes in genotype can result in changes in phenotype.	**1.3** How Living Things Are Alike **8.6** Mutations: Cause and Effect **9.6** Mutated Genes and Their Protein Products **12.4** How Meiosis Introduces Variations in Traits **14.5** Heritable Changes in Chromosome Structure; **14.6** Heritable Changes in Chromosome Number **26** All sections **29.2** Reproductive Structures **41.3** Organs of Sexual Reproduction; **41.4** Reproductive System of Human Females; **41.5** Female Reproductive Cycles; **41.6** Reproductive System of Human Males; **41.7** Bringing Gametes Together **43.2** Behavioral Genetics; **43.7** Mates, Offspring, and Reproductive Success

EK 3.C.2: Biological systems have multiple processes that increase genetic variation.	**12.4** How Meiosis Introduces Variations in Traits **17.2** Individuals Don't Evolve, Populations Do **29.3** Flowers and Their Pollinators; **29.5** Flower Sex **41.3** Organs of Sexual Reproduction; **41.4** Reproductive System of Human Females; **41.5** Female Reproductive Cycles; **41.6** Reproductive System of Human Males; **41.7** Bringing Gametes Together **43.2** Behavioral Genetics; **43.7** Mates, Offspring, and Reproductive Success
EK 3.C.3: Viral replication results in genetic variation, and viral infection can introduce genetic variation into the hosts.	**20.2** Viruses and Viroids; **20.3** Viral Replication; **20.4** Viruses as Human Pathogens
Enduring Understanding 3.D: Cells communicate by generating, transmitting, and receiving chemical signals.	
EK 3.D.1: Cell communication processes share common features that reflect a shared evolutionary history.	**4.11** Cell Surface Specialization **30–34** All sections **35.7** How Muscle Contracts; **35.8** Nervous Control of Muscle Contraction; **35.9** Muscle Metabolism **37** All sections
EK 3.D.2: Cells communicate with each other through direct contact with other cells or from a distance via chemical signaling.	**4.11** Cell Surface Specializations **30–35** All sections **36.4** The Human Heart **37** All sections **38.6** How We Breathe **40.2** Regulating Fluid Volume and Composition; **40.3** The Human Urinary System; **40.4** How Urine Forms; **40.5** Fluid Homeostasis **41.4** Reproductive System of Human Females; **41.5** Female Reproductive Cycles; **41.6** Reproductive System of Human Males **42** All sections
EK 3.D.3: Signal transduction pathways link signal reception with cellular response.	**5.7** A Closer Look at Cell Membranes **30** All sections **31.9** Negative Feedback in Homeostasis **32–34** All sections **35.4** Bone Structure and Function; **35.8** Nervous Control of Muscle Contraction **37** All sections **41.4** Reproductive System of Human Females; **41.5** Female Reproductive Cycles; **41.6** Reproductive System of Human Males
EK 3.D.4: Changes in signal transduction pathways can alter cellular response.	**31.9** Negative Feedback in Homeostasis **32–34** All sections **35.4** Bone Structure and Function; **35.8** Nervous Control of Muscle Contraction **37** All sections **38.6** How We Breathe **41.4** Reproductive System of Human Females; **41.5** Female Reproductive Cycles; **41.6** Reproductive System of Human Males **42.1** Prenatal Problems
Enduring Understanding 3.E: Transmission of information results in changes within and between biological systems.	
EK 3.E.1: Individuals can act on information and communicate it to others.	**29.3** Flowers and Their Pollinators **30.10** Responses to Stress **32.11** The Human Cerebral Cortex; **32.12** Emotion and Memory **33.4** Chemical Senses **34.9** The Adrenal Glands

	41.2 Modes of Animal Reproduction **43** All sections **45.2** What Factors Shape Community Structure?; **45.3** Mutualism; **45.4** Competitive Interactions; **45.5** Predator-Prey Interactions
EK 3.E.2: Animals have nervous systems that detect external and internal signals, transmit and integrate information, and produce responses.	**24.5** Cnidarians—Predators with Stinging Cells; **24.6** Flatworms—Simple Organ Systems; **24.7** Annelids—Segmented Worms; **24.8** Mollusks—Animals with a Mantle; **24.9** Rotifers and Tardigrades—Tiny and Tough; **24.10** Roundworms—Unsegmented Worms that Molt; **24.11** Arthropods—Molting Animals with Jointed Legs; **24.12** Chelicerates—Spiders and Their Relatives; **24.13** Myriapods; **24.14** Crustaceans; **24.15** Insects—Diverse and Abundant; **24.16** The Spiny-Skinned Echinoderms **25.3** Jawless Fishes; **25.5** Modern Jawed Fishes; **25.6** Amphibians—First Tetrapods on Land; **25.8** Nonbird Reptiles; **25.9** Birds—the Feathered Ones; **25.10** Mammals—Milk Makers **31.6** Nervous Tissue **32–33** All sections **35.8** Nervous Control of Muscle Contraction **36.4** The Human Heart **38.6** How We Breathe

Big Idea 4: Biological systems interact, and these systems and their interactions possess complex properties.

Enduring Understanding 4.A: Interactions within biological systems lead to complex properties.

EK 4.A.1: The subcomponents of biological molecules and their sequence determine the properties of that molecule.	**3** All sections **8.3** The Discovery of DNA's Structure **9.4** RNA and the Genetic Code; **9.5** Translation: RNA to Protein; **9.6** Mutated Genes and Their Protein Products **34.3** The Nature of Hormone Action **37.6** Antigen Receptors **39.9** Metabolism of Absorbed Organic Compounds; **39.10** Vitamins, Minerals, and Phytochemicals; **39.11** What Should You Eat?
EK 4.A.2: The structure and function of subcellular components, and their interactions, provide essential cellular processes.	**4** All sections **14.2** Human Chromosomes
EK 4.A.3: Interactions between external stimuli and regulated gene expression result in specialization of cells, tissues, and organs.	**10** All sections **11.6** When Mitosis Becomes Pathological **29, 34, 42** All sections
EK 4.A.4: Organisms exhibit complex properties due to interactions between their constituent parts.	**27.2** The Plant Body; **27.3** Plant Tissues **28–40** All sections **41.3** Organs of Sexual Reproduction; **41.4** Reproductive System of Human Females; **41.5** Female Reproductive Cycles; **41.6** Reproductive System of Human Males **42** All sections
EK 4.A.5: Communities are composed of populations of organisms that interact in complex ways.	**22.9** Angiosperm Diversity and Importance **23.7** Ecological Roles of Fungi **29.3** Flowers and Their Pollinators; **29.7** Fruits **30.10** Responses to Stress

	43–46 All sections **47.5** Biomes; **47.6** Deserts; **47.7** Grasslands; **47.8** Dry Shrublands and Woodlands; **47.9** Broadleaf Forests; **47.10** Coniferous Forests; **47.11** Tundra; **47.12** Freshwater Ecosystems; **47.13** Coastal Ecosystems; **47.14** Coral Reefs; **47.15** The Open Ocean
EK 4.A.6: Interactions among living systems and with their environment result in the movement of matter and energy.	**1.3** How Living Things Are Alike **28.5** Water-Conserving Adaptations of Stems and Leaves **29, 33, 39** All sections **40.2** Regulating Fluid Volume and Composition; **40.5** Fluid Homeostasis; **40.7** Heat Gains and Losses; **40.8** Adaptations to Heat and Cold **44–46** All sections **47.5** Biomes; **47.6** Deserts; **47.7** Grasslands; **47.8** Dry Shrublands and Woodlands; **47.9** Broadleaf Forests; **47.10** Coniferous Forests; **47.11** Tundra; **47.12** Freshwater Ecosystems; **47.13** Coastal Ecosystems; **47.14** Coral Reefs; **47.15** The Open Ocean **48** All sections
Enduring Understanding 4.B: Competition and cooperation are important aspects of biological systems.	
EK 4.B.1: Interactions between molecules affect their structure and function.	**5.4** How Enzymes Work; **5.5** Metabolism—Organized, Enzyme-Mediated Reactions; **5.6** Cofactors in Metabolic Pathways **8.4** Eukaryotic Chromosomes; **8.6** Mutations—Cause and Effect **14.5** Heritable Changes in Chromosome Structure; **14.6** Heritable Changes in Chromosome Number **34** All sections **35.7** How Muscle Contracts **37** All sections **38.2** The Nature of Respiration **39.9** Metabolism of Absorbed Organic Compounds; **39.10** Vitamins, Minerals, and Phytochemicals; **39.11** What Should You Eat?
EK 4.B.2: Cooperative interactions within organisms promote efficiency in the use of energy and matter.	**4** All sections **5.4** How Enzymes Work; **5.5** Metabolism—Organized, Enzyme-Mediated Reactions; **5.6** Cofactors in Metabolic Pathways **20.6** Factors in the Success of Prokaryotes; **20.7** A Sample of Bacterial Diversity; **20.8** Archaea **27–28, 30–31, 37, 39, 41** All sections
EK 4.B.3: Interactions between and within populations influence patterns of species distribution and abundance.	**17.4** Patterns of Natural Selection; **17.5** Directional Selection; **17.6** Stabilizing and Disruptive Selection; **17.7** Fostering Diversity; **17.9** Reproductive Isolation; **17.11** Other Speciation Models **22.9** Angiosperm Diversity and Importance **23.7** Ecological Roles of Fungi **25.11** Modern Mammalian Diversity **29, 43–46** All sections
EK 4.B.4: Distribution of local and global ecosystems changes over time.	**45.8** Ecological Succession; **45.10** Biogeographic Patterns in Community Structure **46.4** Energy Flow; **46.5** Biogeochemical Cycles; **46.6** The Water Cycle; **46.7** The Carbon Cycle; **46.8** Greenhouse Gases and Climate Change; **46.9** Nitrogen Cycle; **46.10** The Phosphorus Cycle **47.2** Global Air Circulation Patterns; **47.3** The Ocean, Landforms, and Climates; **47.4** The El Nino Southern Oscillation **48** All sections

Enduring Understanding 4.C: Naturally occurring diversity among and between components within biological systems affects interactions with the environment.	
EK 4.C.1: Variation in molecular units provides cells with a wider range of functions.	**3, 8–9** All sections **37.6** Antigen Receptors
EK 4.C.2: Environmental factors influence the expression of the genotype in an organism.	**11.6** When Mitosis Becomes Pathological **10.6** Epigenetics **12.1** Why Sex? **13.6** Nature and Nurture **14.5** Heritable Changes in Chromosome Structure; **14.6** Heritable Changes in Chromosome Number **17.4** Patterns of Natural Selection; **17.5** Directional Selection; **17.6** Stabilizing and Disruptive Selection; **17.7** Fostering Diversity **30, 34, 37** All sections **42.1** Prenatal Problems **43.1** Can You Hear Me Now?; **43.2** Behavioral Genetics; **43.4** Environmental Effects on Behavioral Traits
EK 4.C.3: The level of variation in a population affects population dynamics.	**17.4** Patterns of Natural Selection; **17.5** Directional Selection; **17.6** Stabilizing and Disruptive Selection; **17.7** Fostering Diversity **22.9** Angiosperm Diversity and Importance **23.7** Ecological Roles of Fungi **25.11** Modern Mammalian Diversity **44–45** All sections **46.2** The Nature of Ecosystems; **46.3** The Nature of Food Webs
EK 4.C.4: The diversity of species within an ecosystem may influence the stability of the ecosystem.	**21.2** A Collection of Lineages **22.9** Angiosperm Diversity and Importance **23.2** Fungal Traits and Classification; **23.7** Ecological Roles of Fungi **44–45** All sections **46.2** The Nature of Ecosystems; **46.3** The Nature of Food Webs **47.5** Biomes; **47.6** Deserts; **47.7** Grasslands; **47.8** Dry Shrublands and Woodlands; **47.9** Broadleaf Forests; **47.10** Coniferous Forests; **47.11** Tundra; **47.12** Freshwater Ecosystems; **47.13** Coastal Ecosystems; **47.14** Coral Reefs; **47.15** The Open Ocean **48** All sections

Lesson Outline: Unit 1a: Principles of Cellular Life: Biochemistry and Cell Structure

Correlates with Chapters 1–4 in the 15th edition

- **AP® Biology Big Idea 2:** Biological systems utilize free energy and molecular building blocks to grow, reproduce, and maintain dynamic homeostasis.
- **AP® Biology Big Idea 3:** Living systems store, retrieve, transmit, and respond to information essential to life processes.
- **AP® Biology Big Idea 4:** Biological systems interact, and these systems and their interactions possess complex properties.

Brief chapter summaries

Chapter 1 ("Invitation to Biology") is an excellent preview of the entire year—for both the Big Ideas as well as the Science Practices. Figure 1.2 is especially grounding when referred back to repeatedly throughout the year when doing the AP® Investigations.

Chapter 2 ("Life's Chemical Basis") gives a thorough review of the chemistry that most AP® Biology students have already taken and connects it to life. Section 2.5 ("Hydrogen Bonds and Water") is especially important for understanding the rest of the book.

Chapter 3 ("Molecules of Life") gives a concise, yet thorough, treatment of the four main biological molecules. Repeatedly referring back to this chapter throughout the course is in line with AP® Science Practice 7: The student is able to connect and relate knowledge across various scales, concepts, and representations in and across domains.

Chapter 4 ("Cell Structure") discusses the general internal and external structures found in prokaryotic cells and eukaryotic plant and animal cells, as well as gives a preview of the role of the endomembrane system in protein synthesis and transport. This chapter shouldn't be dwelt on too much because many all of the organelles described are referred back to within context in future chapters (e.g., the chloroplast is referred to in depth in Chapter 6)

Chapter 46 ("Ecosystems") links the content in Chapters 1–4 to a larger context (ecosystems). This is an excellent way to bring in AP® Biology Big Idea 2: Energy (biological systems utilize free energy and molecular building blocks to grow, reproduce, and maintain dynamic homeostasis).

CHAPTER 2: LIFE'S CHEMICAL BASIS AND CHAPTER 3: MOLECULES OF LIFE

Chapters 2 and 3 Objectives

- Explain the exchange of matter that occurs between organisms and their environment, specifically carbon, nitrogen, oxygen, phosphorous, and water.
- Discuss how the structure of a water molecule influences its properties.
- Relate the presence and location of a functional group in a biological molecule to its properties.
- Explain the overall structure and function of carbohydrates, proteins, nucleic acids, and lipids and their monomers (structure of *specific* nucleotides, amino acids, lipids, and carbohydrates is not necessary).

AP® is a trademark registered by the College Board, which is not affiliated with, and does not endorse, this product.

- Predict how the surface area to volume ratio of a cell will influence the rate of cellular transport.
- Evaluate how changes in monomer arrangement and sequence, as well as directionality of biological molecules, can influence their functionality.

Enduring Understanding 2.A: Growth, reproduction, and maintenance of the organization of living systems require free energy and matter.

Essential Knowledge 2.A.3: Organisms must exchange matter with the environment to grow, reproduce, and maintain organization.

Suggested Prior Homework: Have students read Chapters 1 through 3 in the 15th edition over the summer. Consider giving a quiz or small test in the first few days of school.

Chapters 2 and 3 Warm Up Questions

1. Describe which biological molecules the different elements for life end up in.
2. Discuss some of the ways organisms use and rely on water and how hydrogen bonding makes that happen.
3. Why is carbon such an important element in biology?
4. Name common sources of carbohydrates, proteins, and fats in our food.
5. How are macromolecules formed? Give an example.
6. What types of chemical reactions (Section 3.3) does our body go through in order to process whole proteins, such as those found in egg whites compared to if we eat simple amino acids?

Enduring Understanding 4.A: Interactions within biological systems lead to complex properties.

Essential Knowledge 4.A.1: The subcomponents of biological molecules and their sequence determine the properties of that molecule.
Starr/Taggart; *Biology: The Unity and Diversity of Life*, 14th edition, Chapter 3.

Chapters 2 and 3 Lesson Opener

Speedy: (10–15 minutes) Ask students to brainstorm about the elements required by living things for survival and which of the four molecules of life those elements will be found in (Chapter 3). This can be completed as a whole class warm up, collaborative pairs, small, or whole group discussion. Once a class list of required elements is agreed upon, ask students to predict the sources of these elements in the environment and how they are acquired by living things (this later connects to the ecology unit and the optional example of digestion).

Extensive: (40–50 minutes) Have students work in small groups (two to three students) to create a list of the key elements needed by living things for survival and which of the four molecules of life those elements will be found in (Chapter 3). Once a list is created, have students construct a large (poster board or bulletin board paper) flow chart that shows the flow of these elements from the environment to living things and vice versa. You can supplement prior knowledge by giving students a series of terms (carbon dioxide, photosynthesis, nitrification, etc.) to aid them in completing the chart. You may choose to allow them to use their textbooks to look up concepts in future chapters, especially Chapter 46. Throughout the unit, have students add on to their boards with more details as they learn them.

This activity reinforces the need for an exchange of matter between living things and their environment and helps to ensure that learning in this unit is rooted to this overarching concept, rather than isolated learning.

Inquiry Opening: Provide students with items that might be used to explore the properties of water (celery, food coloring, paper clips, mesh screen, rubber band, empty beakers, cylinders or cups, pennies, pipettes, etc.). Challenge the students to work in teams to come up with as many methods of demonstrating the various properties of water. Remind students that they must be able to explain how each of their demonstrations shows each property. This makes a great challenge activity, as well as team-building exercise for lab groups.

Chapters 2 and 3 Suggestions for Presenting the Material

- Use simplified cardboard or paper cutouts of amino acids, nucleic acids, monosaccharides, and triglycerides and/or phospholipids. Have students build complex molecules from those building blocks.
- Use graphically rich PowerPoint presentation with embedded video clips.
- Add relevance to your lesson by connecting content to tangible examples such as forensics, medicine, pharmacology, and ecology.
- Encourage the incorporation, restatement, and connection back to the AP® Big Ideas and Enduring Understandings throughout the lesson.
- Quickly review of the use of the periodic table.
- Use molecular models when discussing bonding patterns and structure of water.

Chapters 2 and 3 Common Student Misconceptions

Many students overlook the connection between the structure of a molecule and its function. This unit provides numerous opportunities to highlight this concept including carbon tetravalence and its presence in biological molecules, the polarity of water and its unique properties, the directionality of a DNA molecule and its information, and the levels of structure and conformation of protein molecules.

Chapters 2 and 3 Other Classroom Discussion and Activities

1. Fill a large jar with water, then add salad oil. Shake the bottle, then allow it to sit on the front desk. Ask students to explain what has happened. Add a few drops of methylene blue (a polar dye) and Sudan III fat stain (a nonpolar dye) to the jar and shake. Students will note that the water layer is blue and the oil layer is red; ask them why this is so.
2. Cover a raw chicken egg with vinegar in a small cup or plastic container. After three or four days, the weak acid (vinegar) will eat away the shell, leaving the cell membrane and egg contents intact. With care, one can handle the shell-less egg and illustrate how even weak acids must be buffered by living organisms.
3. Initiate a class discussion on acid rain. What is it? Is acid rain a problem where you live? Ask students to come up with a chemical formula for acid rain. What are some of the effects of this kind of precipitation? Allow students to read an overview of acid rain (e.g., National Geographic: http://environment.nationalgeographic.com/environment/global-warming/acid-rain-overview/) before having a class discussion.
4. Using the names of the active ingredients on an antacid package, explain how they act as *buffers* to stomach acid.
5. Many elements have radioactive isotopes that are useful as tracers in biological systems. Show how $^{14}CO_2$ can be used to follow the fate of carbon as it is incorporated into carbohydrate. In addition, make the connection between thyroid studies and the use of radioactive iodine.
6. Using unknown Macromolecule Lab using glucose test strips, Benedict's solution, Biuret reagent, and Gram's iodine, detect the various molecules of life in common foods.

7. Set up quick demonstration stations of the key properties of water.
8. Use a video clip of a natural disaster or catastrophic event movie to spark a discussion on the importance of water and climatic impact of large-scale changes to our planetary percentage of liquid water.
9. Print structures of the various macromolecule monomers (monosaccharide, amino acid, and nucleotides) on standard computer paper. Make templates of water molecules and have students model dehydration synthesis and hydrolysis by combining and breaking the "paper macromolecules." The end result can be displayed in the classroom as a large-scale macromolecule (e.g., polysaccharide created from removing H and OH from monomers to join into a polymer).
10. Create a concentration game using index cards with various functional groups and common macromolecule structures to encourage identification.
11. Stress that "shape determines function." If a protein does not fold properly, it may not function properly (or at all) in the body. Discuss with students the implications of protein misfolding and how this may also lead to diseases such as Alzheimer's disease, prion disease (e.g., mad cow disease), and even cancer.
12. Your students will recognize the macromolecules in this chapter as major food groups. You can capitalize on this to generate student interest. Have the students research the ways these varying food groups are digested in the human body.
13. Students have been exposed to many words related to the ones in this chapter, whether in print or broadcast media. Use this opportunity to explain complex carbohydrates, polyunsaturated fats, cholesterol, fiber, high-fructose syrup, dextrose, anabolic steroid, and no- or low-carb diets.
14. Protein *primary structure* can be demonstrated by a string of beads or a Christmas tree garland. Individual beads can be colored with felt-tip markers for greater clarity and distinction. Secondary structure (alpha helix) is adequately illustrated by use of a Slinky®. You can demonstrate tertiary structure by *carefully* folding a portion of the "expanded" Slinky. Even quaternary structure can be visualized by using two (or more) Slinkies. Alternatively, show a brief video in class from RicochetScience detailing basic protein structure: http://ricochetscience.com/videos-chemistry-and-proteins/.
15. Much of our DNA sequence is similar to many other organisms, even those distantly related to us. Why might this be the case?
16. Why don't animal cells contain cellulose? Can you think of at least one reason why cellulose in an animal cell could be considered a drawback?
17. Which yields more energy, a pound of carbohydrate or a pound of fat?
18. Approach the study of the various cycles of an ecosystem by first explaining the necessity of that particular component in the environment. Then examine the deleterious effects of too little or too much of that substance.
19. Stress the fact that, in various sections of the chapter, the authors show how humans have altered the natural ecosystems and their functioning.
20. Have some of the students choose a cycle discussed in this chapter and make a poster depicting the stages. Each participant should explain his poster to the class and how human intervention may alter the natural cycle.
21. Set up aquatic ecosystems in the lab and monitor them throughout the semester. Identify the trophic levels of the ecosystem and analyze the cycling of materials and nutrients within it. In what ways are the aquatic ecosystems in the lab similar to, or different from, a real aquatic ecosystem?

22. Test various water supplies for the presence of nitrogen and phosphorus. Are they within the recommended standards? If not, how far do they deviate from a normal level? Use a nursing or anatomy text to determine if an excess of either substance would cause any short-term or long-lasting effects?
23. Examine the composition of several plant and garden fertilizers. What are the major ingredients? What components provide the "miraculous" growth rates?
24. Is it environmentally wise to rely on large quantities of nitrogen-rich fertilizers for crop production? What are some alternatives? Discuss the pros and cons of commercial fertilizers.
25. Investigate the source of your home water supply. If applicable, see if you can find the name of the aquifer that supplies the water.

Case Studies

- National Center for Case Study in Science: http://sciencecases.lib.buffalo.edu/cs/collection
- Atkins or Fadkins
    - o Investigation of low-carbohydrate diets.
- Face the Fats
    - o Clicker case introduces students to the biochemistry of lipids.
- Secret of the Popcorn Popping
    - o Discusses the important role of water in living cellular chemistry.

Chapter 2 Possible Responses to *Critical Thinking* Questions

1. Lead and gold have different numbers of protons. To change the element, one must change the number of protons in each and every atom of the substance. It is impossible to change the number of protons by chemical means. Today physicists have the ability to add or remove protons by extraordinary means, but this was not possible in ancient time. Also, lead is a very stable substance. Enticing lead to give up three protons would require a huge amount of energy, and the cost involved in doing this would far outweigh the profit realized from changing the lead into gold.

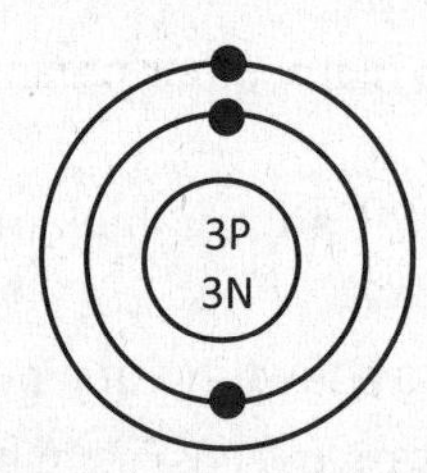

2. Polonium has a half-life of 138 days and decays to form ^{206}Pb (a stable form of lead) after emitting an alpha particle.
3. When an acid is mixed with water, it dissociates and becomes more reactive. Therefore, it is prudent to remove the majority of the acid before washing the area with water.

Chapter 2 Possible Responses to *Data Analysis Activities* Questions

Mercury emissions are a dangerous source of air pollution and are produced by various industries worldwide.

1. Approximately 2,200 tons of mercury was released worldwide in 2006.
2. The source that released the most mercury was the combustion of fossil fuels, and the next greatest emissions resulted from artisanal and small-scale gold production.
3. Asia produced the most mercury from cement production.
4. Approximately 150 tons of mercury was released from South America in 2006 due to gold production.

Chapter 3 Possible Responses to *Critical Thinking* Questions

1. One would accurately predict that the phospholipids would be on the outside of the lipoprotein clump. The phospholipids are well-suited to this function because they are amphipathic (having both a polar and nonpolar end). The polar portion, which is hydrophilic, faces outward so that it is compatible with the surrounding watery environment.
2. Since the literature from experimental studies states that 92.8% of the sucralose passes through the body unchanged, it does not appear that significant quantities of chlorine would be released. In addition, toxicity is unlikely because chlorine is highly reactive and a major element in sodium chloride (NaCl), or salt. Living things are mostly saltwater, so any small amount of chlorine from the sucralose would likely be bound to sodium very quickly. Additionally, it would take massive amounts of sucralose consumed on a daily basis to release enough chlorine to even be noticeable.

Chapter 3 Possible Responses to *Data Analysis Exercise* Questions

The two densities of lipoproteins react differently in the body. High-density lipoproteins (HDL) carry cholesterol to the liver for disposal, but the cholesterol transported by the low-density lipoproteins (LDL) tends to accumulate in arteries as plaque.

1. The LDL level was highest in the group that consumed saturated fats as the main dietary fat.
2. The HDL level was lowest in the group that consumed *trans* fatty acids as the main dietary fat.
3. The group with the highest LDL-to-HDL ratio is the experimental group that consumed *trans* fatty acids as the main dietary fat.
4. The diets, ranked from best to worst, are *cis* fatty acids, saturated fats, and *trans* fatty acids.

Chapter 46 Possible Responses to *Critical Thinking* Questions

1. The vegetable gardens in Maine and Florida would differ greatly based on climate (e.g., rainfall, temperature, latitude, length of growing season, soil content, and perhaps proximity to the ocean), depending on their exact locations. All of these factors would greatly influence what types of crops would be best suited to each region. Florida would have a longer growing season, so I would expect the garden in that area to have a higher annual productivity.
2. The website listed takes you to a site that is sponsored by the United States Geological Survey Department. At this website, you can click on a map to increasingly zoom down to about the county level. It will then give you information about your particular watershed, such as water quality, location of groundwater inventory sites, etc. In addition, it offers detailed definitions of all technical terms for the layperson's interpretation.
3. One would not be able to test the air in bubbles in frozen ice for phosphorus, because phosphorus does not occur naturally in that form. The best way to check phosphorus levels from the past would be to test the levels in sedimentary rock, since this is the major reservoir for phosphorus.
4. If you increased the number of nitrogen-fixing bacteria in an aquatic ecosystem, it could result in an algae bloom and eutrophication. The additional nitrogen would encourage an overgrowth of algae, which may cause a hypoxic atmosphere for the animals that reside there. This eutrophication would lead to an increase in carbon production and an increase in carbon accumulation.
5. If the mycorrhizal fungi were used to provide plants with phosphorus and nitrogen, the risk of eutrophication would decrease dramatically. The runoff from the fields would not contain the excessive amounts of nutrients that would be seen in an area exposed to large amounts of fertilizers. Some research should be performed before such a widespread plan is initiated to ensure that the fungi in large amounts would not cause any significant disruption in the overall balance of nature.

Chapter 46 Possible Responses to *Data Analysis Activities* Questions

Scientists are attempting to show that human activities have impacted the amount of carbon dioxide in the atmosphere. In Antarctica, there is thick ice that can be used to measure the carbon dioxide level over time.

1. The highest carbon dioxide level between 400,000 and 0 A.D. was 300 ppm.
2. During this same time period, the carbon dioxide level never reached the level attained in 1980.
3. The trend in the carbon dioxide level for the 800 years prior to the advent of the industrial revolution was fairly constant, with some slight variations. After 1800, the carbon dioxide level began to rise dramatically.
4. The rise in the carbon dioxide level between 1980 and 2013 was much higher than the rise between 1800 and 1975.

CHAPTER 4: CELLS

Chapter 4 Objectives

- Explain the limits on cell size.
- Compare and contrast prokaryotic and eukaryotic cells.
- Identify cell type from given list of structures present.
- Discuss the link between the structure and function of cells using examples such as nerve, blood, or sex cells.
- Discuss how compartmentalization influenced specialization and complexity.
- Explain how the evolution of the cell theory represents scientific process.
- Discuss how cell compartmentalization influences specialization and complexity.

Enduring Understanding 2.B: Growth, reproduction, and dynamic homeostasis require that cells create and maintain internal environments that are different from their external environments.

Essential Knowledge 2.B.3: Eukaryotic cells maintain internal membranes that partition the cell into specialized regions.

Essential Knowledge 4.B.2: Cooperative interactions within organisms promote efficiency in the use of energy and matter.

Enduring Understanding 4.A: Interactions within biological systems lead to complex properties.

Essential Knowledge 4.A.2: The structure and function of subcellular components, and their interactions, provide essential cellular processes.

Chapter 4 Warm Up Questions

1. How is the structure of a cell related to its function?
2. Give two examples of how the structure of a cell is related to its function.
3. How are cells organized?
4. What benefit does multicellularity provide eukaryotes?
5. How is the function of the chloroplast and mitochondria specialized within a cell?
6. Name five cellular structures in eukaryotic cells.

Chapter 4 Lesson Opener

1. **Speedy:** (10–15 minutes) Show students pictures of four to five different types of cells (nerve, blood, sperm, plant, and bacteria), and ask them to identify structural differences and brainstorm the reasons for these differences. Draw a basic timeline for the cell theory on the board (beginning with the invention of the microscope), and ask students to address the implication of each subsequent discovery on the birth of cell biology.
2. **Extensive:** (40–50 minutes) To introduce the cell cycle and the need for regulation, show a video clip of cells dividing uncontrollably (cancer) and a clip of normal cell growth. Visit www.youtube.com—*Mitosis, Cancer Cells*.
3. Have students look at various types of cells under the microscope. If prepared slides are not available, then the teacher can have the students make slides of their own cheek cells by scraping the inside of the cheek with a toothpick and then placing the toothpick in a drop of saline solution on a slide. Have students add one drop of iodine and put a cover slip on the slide. Look at the epithelial cells at 400×. Students could also look at a leaf of *Elodea* (aquatic plant available at pet stores). Have students make a wet mount with the leaf and look at it under 400×. Have students note the shape of the cell and the chloroplast. Students can add salt water and observe what happens to the cell membrane.
4. Read about Henrietta Lack's "Immortal Cells" at https://www.smithsonianmag.com/science-nature/henrietta-lacks-immortal-cells-6421299/. Discuss the ethics involved in taking someone's cells without their permission.

Chapter 4 Suggestions for Presenting the Material

- For many readers, Chapter 4 represents the true entry into the realm of biology. A discussion of the cell is fundamental to all future lectures.
- Use the video *The Inner Life of the Cell* to enable students to visualize the cell components in motion. Refer back to this video throughout the course. https://www.youtube.com/watch?v=wJyUtbn0O5Y
- Often students entering into the healthcare fields have a difficult time understanding the relevance of cellular physiology to their studies. Be sure to stress that some diseases, such as Tay-Sachs disease, occur on the cellular level.
- This is also an excellent time to review the use of the word *theory* as explained in Chapter 1.
- A clear distinction between prokaryotic and eukaryotic cells should be made (see the Enrichment section later in the text for visual aid suggestion). Additional details on prokaryotes are contained in Chapter 21.
- Although the descriptions and diagrams of the cell organelles occupy only a small number of textbook pages, it is best to proceed carefully and deliberately. There is a dizzying array of unfamiliar terms here.
- It's helpful to make the analogy that organelles are like the different departments/agencies within a town. Not all the organelles will fit into this analogy, but it makes the point that each organelle has a function in the overall workings of the cell as a whole.
- When describing each cell structure, a visual representation of some type should be constantly in view of the students. Each time a new cell part is introduced, Figure 4.11a or 4.11b should be shown for reference purposes to help students visualize where the organelle belongs within the cell.
- Stress the fact that several cell parts are so complex in function that greater detail will follow in future lectures. Examples and chapters are cell membranes in Chapter 5, chloroplasts in Chapter 7, mitochondria in Chapter 8, and nucleus and chromosomes in Unit II.

Chapter 4 Classroom and Laboratory Enrichment

- **AP® Investigation #4: Diffusion and Osmosis, Procedure 1: Cell Size Lab**
- In the lab, students can view what Robert Hooke saw by looking at cork cells under a microscope. Explain that what they are seeing is the dead cell walls of the cork plant. They can also look at a small *Elodea* leaf to view the chloroplasts, nucleus, and a few other plant cell features. Place the leaf in a small drop of distilled water on a microscope slide. Cover it with a cover slip. Now prepare a second slide, only this time mounting an *Elodea* leaf in a drop of 10% NaCl solution. Compare the cells of the second slide to those of the first slide.
- Have the students examine common representative cells with different magnification levels. Which magnification strength is best for viewing the cellular structures? Do the students know how to calculate the magnification of a light microscope?
- Use sketches or models drawn to scale to demonstrate the size difference between prokaryotic and eukaryotic cells.
- Show a picture of any cell. Ask if the cell is prokaryotic or eukaryotic. Is it an animal cell? A plant cell? Some other type of cell? How can they tell?
- Ask students to match "organelle" with "cellular task" at the board or on an overhead.
- Most departments possess some type of cell model. These are especially helpful in perception of the three-dimensional aspects of cell structure. They can also be useful in oral quizzing.
- Have the students match each organelle to its function in plant and/or animal cells. For those organelles found only in a plant or animal cell, discuss why that organelle is only necessary in that one type of cell. Direct students to explore the inside of a cell using an interactive cell from University of Utah available at http://learn.genetics.utah.edu/content/cells/insideacell/.

Chapter 4 Classroom Discussion Ideas

- Why are some strains of *E. coli* helpful while others make us sick?
- How can people protect themselves from becoming sick with *E. coli*?
- What is "finely textured beef" and is there any controversy over this type of beef?
- What are some organelles that contain internal compartmentalizations? How do internal compartments assist in the functioning of the organelle?
- List the tasks that a cell must perform.
- Why are there no unicellular creatures one foot in diameter?
- Why do bacteria have ribosomes when they lack other organelles?
- What is the difference between scanning electron microscopy and transmission electron microscopy?
- Why do you think most plant cells have a central vacuole while animal cells lack this organelle?
- What is the significance of the word *theory* in reference to the basic properties of the cell?
- Does the cytoplasm have any functions of its own, or is it just a "filler" matrix in which other organelles float? What are the major chemical components of cytoplasm?
- In which cells might you find a high concentration of peroxisomes? What would be the function of the peroxisomes?
- Show the class some slides of representative cells. Can the class determine the cell type by identifying the cellular components it contains? Often the presence of large quantities of particular structures in a cell can indicate its function. For example, show a slide of a sperm cell. The students may notice a large amount of mitochondria in the midpiece area. This indicates that this cell has a high energy demand. How does this relate to the presence of a flagellum? Next, look at a slide of a mature red

blood cell. The students may be surprised to see that it does not contain a nucleus. This explains the fact that a red blood cell cannot divide. Explain that the function of a red blood cell is to carry oxygen, so it expels the nucleus to allow for more room for the transportation of oxygen.

- Why is the term *nucleus* used to describe the center of an atom and the organelle at the center of the cell when these are such different entities?
- In measurement of length, what are the largest cells (when mature) in the human body? What fundamental property of all cells is denied to these?
- What feature makes a eukaryotic cell a "true" cell?
- Why aren't there one-celled organisms as big as beach balls? What modifications would have to be made in distribution within the cell for such an event to occur?
- What organelle could be compared to the "control center" of an assembly line in a factory?
- Describe the interrelationship(s) of the individual members comprising the *endomembrane system* (see Figure 4.15).

Chapter 4 Homework Extension Questions

1. Discuss the benefits of compartmentalization.
2. Explain the factors that influence cell size.
3. Identify and discuss the differences between eukaryotic and prokaryotic cells.
4. Describe how the interaction in a biofilm improves efficiency and utilization of energy.
5. State the benefits of organelle specialization in eukaryotes.
6. Explain why the specialization found within the lungs is critical to efficient gas exchange.

Chapter 4 Possible Responses to *Critical Thinking* Questions

1. The idea of a gigantic cell of any kind, or a gigantic creature of any kind, has fascinated Hollywood moviemakers for decades. Moviegoers love this sort of thing. But wet-blanket scientists have to spoil the fun by pointing out the impossibility of gigantic cells based on that surface-to-volume ratio thing. Simply put, a huge cell would have so much volume that the distance from the deepest reaches of the interior of the cell would be too far for nutrients and oxygen to diffuse in, and for carbon dioxide and metabolic waste to diffuse out, not to mention the woefully inadequate surface area for all that diffusion to take place. As for encountering a giant amoeba in space, it would be impossible. You have already seen they can't exist and even if one did, it would have to survive in the vacuum of space!
2. The secondary plant cell wall, which is often deposited inside the primary cell wall as a cell matures, sometimes has a composition nearly identical to that of the earlier-developed wall. Because additional substances, especially lignin, are found in the secondary wall, it must be deposited from the inside where it is produced by the cell. The microfibrils of the secondary cell wall are aligned mostly in the same direction, and communication is possible by pits in the secondary cell wall that allow plasmodesmata to connect to the cell.
3. *Euglena* is a eukaryotic cell, which one knows from the presence of the distinct nucleus. These single-celled protists may have plant or animal characteristics. The electron micrograph on page 75 is of a *Euglena* that performs plantlike functions, since there are chloroplasts evident. Other structures that can be identified in the image are flagellum, vacuoles, and the cell membrane.

Chapter 4 Possible Responses to *Data Analysis Activities* Questions

In a normal flagellum, one sees a pattern of nine pairs of microtubules arranged in a circle, with the protein dynein located between each pair. Abnormal dynein would affect the movement of both flagella and cilia. This would result in sperm that could not swim normally to perform fertilization and cilia that would have trouble sweeping the respiratory tract free of debris.

AP® Practice Essays

1. All life on Earth is composed of cells. They vary in size, shape, and complexity.
   a. Describe THREE features common to all cells.
   b. Discuss the major factor that limits the size that a typical cell can attain.
2. Water is essential to life.
   a. Describe the fundamental structure of a water molecule.
   b. Describe how water's structure is related to its properties.
3. Discuss the following:
   a. The levels of organization of a protein.
   b. What could cause destruction of a protein? How is this related to its function?
   c. How does DNA determine a protein's structure?
4. Nutrient cycles describe how water, phosphorus, nitrogen, and carbon cycle between and within the biotic and abiotic environments.
   a. Describe the processes that occur for a carbon atom to cycle from an amino acid in a leaf cell to a protein that eventually gets eaten by a caterpillar who eventually breathes it out only to get taken up by a leaf again.
   b. Describe one way in which water cycling through the water cycle can influence nitrogen in the nitrogen cycle.
5. All life on earth is composed of cells. They vary in size, shape, and complexity.
   a. Describe THREE similarities common in all cells.
   b. Discuss the major factor that limits the size that a typical cell can attain.
   c. Describe TWO levels of organization associated with eukaryotic cells. Explain how this organization allows these cells to function more efficiently.

Lesson Outline: Unit 1b: Principles of Cellular Life: Cellular Metabolism

Correlates with 14th edition book, Chapters 5–7, with connections to Chapter 46 and others in the 15th edition

- **AP® Biology Big Idea 2:** Biological systems utilize free energy and molecular building blocks to grow, reproduce, and maintain dynamic homeostasis.
- **AP® Biology Big Idea 3:** Living systems store, retrieve, transmit, and respond to information essential to life processes.
- **AP® Biology Big Idea 4:** Biological systems interact, and these systems and their interactions possess complex properties.

Brief chapter summaries

Chapter 5 ("Ground Rules of Metabolism") correlates with AP® Big Idea 2 very nicely. The first six sections correlate to how energy is harnessed, transformed, and used in cells and living systems. The last four sections describe how cells can transfer materials (nutrients and wastes) within cells and across the cell membrane.

Chapter 6 ("Where It Starts—Photosynthesis") describes how the majority of the energy in living systems gets captured by the sun.

Chapter 7 ("How Cells Release Chemical Energy") then describes how the energy captured by photosynthesis into organic compounds can then be utilized via cellular respiration.

Chapter 46 ("Ecosystems") also links the content in Chapters 5–7 to a larger context (ecosystems). This is an excellent way to bring in AP® Biology Big Idea 2: Energy (biological systems utilize free energy and molecular building blocks to grow, reproduce, and maintain dynamic homeostasis).

CHAPTER 5, Sections 5.1–5.6: ENERGY TRANSFER

Chapter 5, Sections 5.1–5.6 Objectives

- Describe the physical and biochemical characteristics of an enzyme.
- Explain how the structure of an enzyme is related to its function.
- Discuss the mechanisms of enzymatic control.
- Evaluate how changes in free energy affect organisms and their environment.
- Identify reaction type, activation energy, and Δ G when given a graph.
- Discuss how photosynthesis and cell respiration are connected in the global energy cycle.

Connect the contents of Chapter 5, Sections 5.1–5.6, to energy transfer in ecosystems (Chapter 46), maintaining body temperature in ectotherms and endotherms (Chapter 25, Section 25.7), and reproduction of organisms (Chapter 41).

Chapter 5, Sections 5.1–5.6 Lesson Opener

Speedy: (10–15 minutes) Create an energy diagram connecting the molecules involved in cell respiration and photosynthesis, including all reactants and products. Through discussion, add exergonic and endergonic to the diagram, heat loss, and the one-directional flow of energy. This helps students to focus on the big picture ideas that connect the global energy cycle. This is a useful way to assess prior knowledge before the unit to identify strengths and weaknesses in background information.

AP® is a trademark registered by the College Board, which is not affiliated with, and does not endorse, this product.

Extensive: (40–50 minutes) Ask students to explore the Biosphere 2 website (www.b2science.org). Facilitate a group discussion, or assign two student facilitators, on the resource requirements needed to sustain a facility like a biosphere. Guide the discussion to highlight energy requirements and the interdependence of organisms. As an extension activity after the unit or elaboration assignment during the unit, students can design a similar facility to be placed on the moon, emphasizing resource cycling, energy use, and acquisition and the role of organisms.

Enduring Understanding 4.B: Competition and cooperation are important aspects of biological systems.

Essential Knowledge 4.B.1: Interactions between molecules affect their structure and function.

Enduring Understanding 2.A: Growth, reproduction, and maintenance of the organization of living systems require free energy and matter.

Essential Knowledge 2.A.1: All living systems require constant input of free energy.

Warm Up Questions

1. Describe the motion of molecules at cold and warm temperatures.
2. How is energy used in living things?
3. Identify two endothermic "warm-blooded" and two ectothermic "cold-blooded" organisms.
4. Why is the efficiency of metabolic pathways critical?
5. How is the shape of a protein vital to its function?
6. What does it mean to be lactose intolerant?

Chapter 5, Sections 5.1–5.6 Suggestions for Presenting the Material Relating to Energy Transfer and Enzymes

- Because students may be unfamiliar with the first and second laws of thermodynamics, it is important to distinguish clearly between the two laws; emphasize the central role of the sun in sustaining life on Earth.
- Acknowledge that this chapter speaks in *generalities* and defines terms that will be used to describe *specific* metabolic reactions in subsequent chapters.
- After presentation of the various capabilities of enzymes, students may think of them as "miracle workers." Remind the students that these are nonliving molecules—albeit, amazing ones. Also, emphasize the limitations and vulnerability of enzymes, including causes and effects of denaturation.
- Tie endergonic reactions to energy storage and phosphorylation of ADP to ATP, exergonic reactions to ATP breaking the outer phosphate bond to form ADP and release energy.
- Encourage students to bring their textbook to class and refer frequently to the excellent figures in it.
- This chapter provides a good opportunity to introduce the idea that oxidations release energy, reductions require energy, and NADH and NADPH contain more potential energy than their oxidized counterparts. Students should have some idea of oxidation/reduction from chemistry classes, but it may help them remember if you teach this mnemonic acronym: LEO says GER (Lose Electrons Oxidation; Gain Electrons Reduction).

Chapter 5, Sections 5.1–5.6 Classroom and Laboratory Enrichment

- **AP® Investigation #13: Enzyme Activity**
- The action of an enzyme (salivary amylase) can be easily demonstrated by the following procedure:
  a. Prepare a 6% starch solution in water and confirm its identity by a spot plate test with iodine solution (produces blue-black color).
  b. Collect saliva from a volunteer by having the person chew a small piece of Parafilm and expectorate into a test tube.
  c. Place diluted saliva and the starch solution in a test tube and mix.
  d. At suitable intervals, remove samples of the digestion mixture and test with iodine on the spot plate (lack of dark color indicates conversion of starch to maltose).
  e. Variations can include heating the saliva to destroy the enzyme or adding acid or alkali.
- Show students a video depicting the role and function of enzymes. For example, "Enzyme Catalysis" available at http://www.bozemanscience.com/science-videos/2012/3/19/ap-biology-lab-2-enzyme-catalysis.html.
- Toothpickase activity (modeling enzyme regulation)—biological supply companies or online.
- Experiment testing the activity of a lactose intolerance medication on the digestion of lactose in various types of milk.
- Use tennis balls to represent electrons. Ask five students to line up in front of the class and then pass the tennis balls, one at a time, from one student to the next. As a student receives a tennis ball (electron), have them say "reduced." As a student passes an electron, have them say "oxidized." With a little practice, the point becomes obvious to the participating students and the class that movement of electrons defines redox reactions.
- Demonstrate the two models of enzyme–substrate interactions in the following ways:
  a. *Rigid "lock and key" model:* Use preschool-size jigsaw puzzle pieces or giant-size Lego blocks.
  b. *Induced-fit model:* Use a flexible fabric or latex glove to show how the insertion of a hand (substrate) induces change in the shape of the glove (active site).
- You can also demonstrate the action of amylase breaking down starches into sugars as the first step of digestion in your mouth. Give each student (or selected volunteers) a saltine cracker and ask them to chew it without swallowing it, keeping it in their mouth as saliva does its work. Ask them what they can taste after it has been in their mouths for a bit—it should begin to taste sweet. Amylase has broken the starch in the cracker into sugar. Students can easily begin to think about the action of enzymes with this demonstration.
- Show a video clip from *Lorenzo's Oil* to demonstrate the effects of a disorder caused by an error in the function of an enzyme, and discuss the importance of structure to function. Draw a sample metabolic pathway on the board and discuss the overall effect on the pathway if enzymes at various points are no longer functioning or are altered by feedback inhibition.
- Use the case study "Why Is Patrick Paralyzed" by National Center for Case Study Teaching in Science to connect the energy concepts of enzymes, regulation of metabolic pathways, and aerobic respiration in a relevant medical dilemma.

Chapter 5, Sections 5.1–5.6 Homework Extension Questions

1. Describe the interaction between a substrate and an enzyme.
2. Discuss how the activity of enzymes is regulated.
3. What environmental conditions can affect the rate of an enzymatic reaction?
4. Explain how energy coupling is used in metabolism.
5. How would changes in free energy affect cells, organisms, and ecosystems?
6. Differentiate between the metabolism of endotherms and ectotherms.

CHAPTER 5, SECTIONS 5.7–5.10: TRANSFER OF MATTER, MEMBRANE STRUCTURE, AND FUNCTION

Objectives

- Distinguish between the types of cellular transport.
- Explain the role of the cell membrane in cellular activities.
- Describe the fluid mosaic model of the membrane.
- Relate the structure of the membrane to its function.
- Explain how tonicity and osmotic pressure affect cells (with and without walls).

Enduring Understanding 2.A: Growth, reproduction, and maintenance of the organization of living systems require free energy and matter.

Essential Knowledge 2.A.1: All living systems require constant input of free energy.

Essential Knowledge 2.A.3: Organisms must exchange matter with the environment to grow, reproduce, and maintain organization.

Enduring Understanding 2.B: Growth, reproduction, and dynamic homeostasis require that cells create and maintain internal environments that are different from their external environments.

Essential Knowledge 2.B.1: Cell membranes are selectively permeable due to their structure.

Essential Knowledge 2.B.2: Growth and dynamic homeostasis are maintained by the constant movement of molecules across the membrane.

Warm Up Questions

1. What substances does a cell need to exchange with its environment?
2. What is the membrane's role in homeostasis?
3. What happens when perfume is sprayed at the front of the classroom over time?
4. Predict the characteristics required of the cell membrane to support cellular survival.
5. Describe the basic structure of a lipid.
6. Describe the basic structure of a phospholipid.

Chapter 5, Sections 5.7–5.10 Lesson Opener

Speedy: (10–15 minutes) **AP® Investigation #4, Procedure 2 (Modeling Diffusion and Osmosis)** is a good lab to use as an introductory demonstration. If you are looking for something super speedy, just show this YouTube video (https://www.youtube.com/watch?v=2Th0PuORsWY), being sure to explain and/or let students ask questions.

Extensive: (40–50 minutes) **AP® Investigation #4, Procedure 2 (Modeling Diffusion and Osmosis)** can also be done as an inquiry lab. Just be sure that the students know before designing their experiments how to test for the presence of each molecule in the bag and/or beaker following the incubation.

Chapter 5, Sections 5.7–5.10 Suggestions for Presenting the Material

- Create an image-based slide presentation to reinforce concepts discussed in class.
- Use micrographs of the cell membrane to illustrate the structure of the bilayer.
- Show an animated clip showing transport through the membrane via transport proteins.

- Use a dialysis tubing demonstration of osmosis before presenting tonicity, and use visual examples from this demonstration to explain osmotic pressure.
- Have students act out passive and active transport. Students can narrate the movement in small groups.
- Show animated cell cycle diagrams or short video segments of each of the phases of the cell cycle.
- During lessons on cells and the cell membrane, stress the relationship between structure and function.
- Don't assume students know, or remember, the definitions of the terms *hydrophilic* and *hydrophobic*.
- Carefully distinguish between which portion of the plasma membrane is the "fluid" (that is, lipid bilayer) and which is the "mosaic" (that is, proteins).
- The various methods by which molecules move, either through space or through membranes, can be confusing to students because of the subtle differences that distinguish each method. Perhaps begin with general, non-membrane-associated phenomena such as diffusion. Then proceed to membrane-associated mechanisms such as osmosis, facilitated diffusion, and active transport. (See the Enrichment section later in the text for ideas on introducing diffusion.)
- Use the analogy of canoeing down a river to illustrate passive transport. One can sit back and let the current move the canoe down the river and get to the desired destination without using energy to paddle the canoe. Likewise, if one is traveling up-river, it takes a lot of paddling energy to move the canoe against the current to the desired destination. Active transport is like trying to paddle up-river.
- Having various photographs of endocytosis and exocytosis from actual cells or protists can help students begin to understand the complexities involved in the processes. Include photographs of receptor-mediated processes and explain how some pathogens are able to "trick" our cell membrane receptors in order to "dock" and gain access to our cells.
- Teach the students the prefix "hyper" means "more" and "tonic" refers to solutes. So a hypertonic solution has more solutes and, therefore due to the spheres of hydration around each solute, less *free* water and, therefore, water will flow to the area with less *free* water. Conversely, the prefix "hypo" means "less," so a hypotonic solution will have fewer solutes and, therefore, a higher free water concentration.
- Use the adage, "Water follows salt" to help the students determine where water moves in isotonic, hypertonic, and hypotonic solutions.

Chapter 5, Sections 5.7–5.10 Classroom Discussion and Activities

- **AP® Investigation #4 Diffusion and Osmosis, Procedure 3: Modeling Osmosis in Living Cells**
- Discuss how new discoveries in science can be incorporated into existing knowledge using Peter Agre's discovery of aquaporins and his subsequent Nobel Prize. Use the information at nobelprize.org to spark this discussion on the process of science and membrane structure.
- Discuss how our understanding of the structure of the membrane evolved into the current fluid mosaic model.
- AP® Biology recommended Lab: Diffusion and Osmosis.
- AP® Biology recommended Lab: Mitosis/Meiosis.
- Kidney Dialysis Lab available from various biological supply companies in kit form using unfiltered simulated blood—link to Chapter 40!
- Bacterial identification and Gram staining lab—students can use dichotomous key to identify bacteria using their structure.
- Investigate the evolutionary connection between prokaryotes and eukaryotes using cell structure.
- Discuss the implication of salt deposits on irrigated crops, after reviewing osmosis and tonicity.

Chapter 5, Sections 5.7–5.10 Homework Extension Questions

1. Discuss specific methods used to maintain homeostasis through transport.
2. Use the process of osmosis to explain what would happen to a red blood cell submerged in a beaker of distilled water.
3. Give an example of a cell using exocytosis to maintain homeostasis.
4. How is the membrane structure related to its function?
5. How do phospholipids and transport proteins aid in and control cellular transport?
6. Contrast the structure and function of the cell wall and membrane.

Chapter 5 Possible Responses to *Critical Thinking* Questions

1. The quantity of energy in the universe comes from the first law, which states that matter (energy) is neither created nor destroyed but remains constant, merely changing forms. You can't gain or "win" energy because there is a finite amount in the universe. The second law deals with the quality of that energy since most of the energy is lost as heat in the transformations. The quality of energy at the beginning of a reaction is not the same as at the end. Entropy measures how much and how far the form of energy disperses after an energy change. Energy tends to flow from concentrated to less concentrated forms spontaneously. So, one cannot "break even" and end up with the same amount of usable energy at the end of energy conversions; much is lost to entropy.
2. Diffusion increases entropy, because the substance goes from a high concentration to a lower concentration. Therefore, this process can be seen as dispersing the energy more.
3. Glucose is transported into the cytoplasm from the extracellular environment by facilitated diffusion. This is an example of passive transport, during which solutes move along their concentration gradient. The entry of glucose into a cell is therefore regulated by the difference in the intracellular and extracellular concentrations; when there is excess glucose outside the cell, it will diffuse into the cytoplasm.
4. Trypsin is an enzyme that is typically produced in the pancreas. Its function is to finish the digestion of proteins that begins in the stomach. If you take food with trypsin, the digestion of proteins will instead begin in your mouth.
5. The gas is oxygen gas (O_2): $H_2O_2 + H_2O_2 +$ catalase $\rightarrow 2H_2O + O_2 +$ catalase

Chapter 5 Possible Responses to *Data Analysis Activities* Questions

Some species of *Ferroplasma* are able to survive in, and even prefer, an acidic environment.

1. The dashed line indicates the optimum enzyme activity for each enzyme.
2. All of the four enzymes have an optimum pH of less than 5. Only two of the enzymes show a significant function at a pH of 5.
3. The optimum pH for carboxylesterase is around 1.5.

CHAPTERS 6 AND 7: PHOTOSYNTHESIS AND CELLULAR RESPIRATION

Chapters 6 and 7 Objectives

- Describe the ATP/ADP cycle.
- Explain the role of water in photosynthesis.
- Summarize the light reactions and Calvin cycle.
- Explain the connection between the light-dependent reactions and Calvin cycle in photosynthesis.
- Discuss how environmental factors (temperature, carbon dioxide, and oxygen concentration) affect the rate of photosynthesis.

- Identify how energy is transferred in energy processes such as cell respiration and photosynthesis.
- Explain how the electron transport chain and chemiosmosis work together to generate ATP.
- Discuss the goal of fermentation in anaerobic respiration.
- Describe the structure of the mitochondria and its role in aerobic respiration.
- Summarize glycolysis, the Krebs cycle, and the electron transport chain.
- Explain the role of oxygen in the generation of ATP.
- Compare and contrast the process of chemiosmosis in photosynthesis and cell respiration.
- Discuss how the structure of the membrane influences the production of ATP.

Enduring Understanding 2.A: Growth, reproduction, and maintenance of the organization of living systems require free energy and matter.

Essential Knowledge 2.A.1: All living systems require constant input of free energy.

Essential Knowledge 2.A.2: Organisms capture and store free energy for use in biological processes.

Enduring Understanding 4.B: Competition and cooperation are important aspects of biological systems.

Essential Knowledge 4.B.2: Cooperative interactions within organisms promote efficiency in the use of energy and matter.

Chapters 6 and 7 Warm Up Questions

1. Why are most parts of plants green?
2. How do autotrophs and heterotrophs each acquire the energy they need for cellular processes?
3. Compare and contrast the main function of both chloroplasts and mitochondria.

Chapters 6 and 7 Lesson Opener

Speedy: (10–15 minutes) Create an energy diagram connecting the molecules involved in cell respiration and photosynthesis, including all reactants and products. Through discussion, add exergonic and endergonic to the diagram, heat loss, and the one-directional flow of energy. This helps students to focus on the big picture ideas that connect the global energy cycle. This is a useful way to assess prior knowledge before the unit to identify strengths and weaknesses in background information.

Extensive: (40–50 minutes) Ask students to explore the Biosphere 2 website (www.b2science.org). Facilitate a group discussion, or assign two student facilitators, on the resource requirements needed to sustain a facility like a biosphere. Guide the discussion to highlight energy requirements and the interdependence of organisms. As an extension activity after the unit or elaboration assignment during the unit, students can design a similar facility to be placed on the moon, emphasizing resource cycling, energy use, and acquisition and the role of organisms.

CHAPTER 6: WHERE IT STARTS—PHOTOSYNTHESIS

Chapter 6 Suggestions for Presenting the Material

- Many students think that the only autotrophic organisms are those with green pigments, which are, of course, capable of using light for food manufacture. Point out the existence of the many *chemoautotrophs*, which derive their energy from chemicals. Use deep-sea thermal vents as an example of where chemoautotrophs can be found.

- A discussion of the reason why colored objects appear the color they do may facilitate the students' understanding of why, for example, green light is ineffective for photosynthesis (not absorbed).
- ATP and NADPH were introduced earlier as carrier molecules. Their connection between the light-dependent and light-independent reactions is an excellent opportunity for reinforcement of these molecules as bridges.
- Even though *cyclic photophosphorylation* is presented *prior* to the noncyclic pathway in the text, note that this is done in deference to its position as the first to evolve. Point out that most existing plants use the noncyclic pathway.
- Tell the students to master each phase of photosynthesis before going to the next. Compartmentalizing the overall process will make the material less overwhelming to students and allow them to "connect the dots" when each phase is mastered.
- Although the diagram of the Calvin-Benson cycle (Figure 6.10) is intimidating, the most important features are the entry of *carbon dioxide* and the production of the sugar phosphate, driven by *ATP* and *NADPH* from the light-dependent reactions.
- Bring in plant examples of C3, C4, and CAM plants. Note which plants are typical in the area where you are teaching and how that ties to climate. Have students explain how the features of C3, C4, and CAM plants are adapted to the environment in which they are found.
- Illustrate the mechanism by which guard cells open and close the stomata to regulate moisture loss.

Chapter 6 Classroom and Laboratory Enrichment

- **AP® Investigation #5: Photosynthesis This lab may be done effectively during many different parts of the course: During the photosynthesis unit, during Unit 5 (Plants) along with AP® Investigation #6: Cellular Respiration**, or in the Ecology unit (Chapters 44–48).
- You can demonstrate the production of oxygen by plants with the following:
  a. Place *Elodea* (an aquarium plant) in a bowl and expose it to bright light.
  b. Invert a test tube over the plant and collect the bubbles.
  c. Remove the tube and immediately thrust a glowing wood splint into the tube.
  d. Result: The splint burns brightly in the high-oxygen air.
- Separate the pigments in green leaves by using paper chromatography. (Consult a botany or biology laboratory manual for the correct procedure.)
- When teaching in the fall, bring in leaves of different fall colors. Have the students identify the pigments that remain once chlorophyll has been degraded. Have the student speculate as to whether these pigments are present when leaves are green in the summer? If so, why can't you see them?
- Many students have never seen the action of a prism of separating white light into its component colors; a demonstration would most likely be appreciated.
- Use a light microscope to view living *Euglena*. Note the chlorophyll in this photosynthetic protist.
- Cut out separated leaf pigments from a paper chromatogram and elute each pigment from the paper with a small amount of alcohol. Using a spectrometer, determine the absorption spectrum for each pigment, and graph the results. (Consult a biology or botany laboratory manual for the procedure.)
- Remove two leaves from a plant and put each in a separate flask. Add sodium hydrogen carbonate to one flask and potassium hydroxide solution to the other. Seal each flask so it is air tight. Wait several days and examine the leaves. The leaf with the sodium hydrogen carbonate looks okay, but the leaf with the potassium hydroxide solution appears to be very weak. This is because the potassium hydroxide solution removed all of the carbon dioxide, so photosynthesis could not occur.
- Have students make large posters of both phases of photosynthesis and explain the processes to the class.

Chapter 6 Classroom Discussion Ideas

- C3 plants grow very quickly, whereas C4 plants grow slowly. CAM plants grow even more slowly still. Discuss the trade-off of slower growth for high temperature and drought tolerance in these three categories of plant.
- How might altitude affect what kind of plants predominate an area? Look into the plants found at each level of Mt. Washington in New Hampshire. You will see a vast difference in the plants that exist at each altitude level of this tall mountain!
- In what ways could the "greenhouse effect" hurt agriculture? In what ways could it possibly *help*, especially in Canada and the countries of the former Soviet Union?
- Suppose you could purchase light bulbs that emitted only certain wavelengths of visible light. What wavelengths would promote the most photosynthesis? The least?
- Assume you have supernatural powers and can stop and start the two sets of photosynthesis reactions. Will stopping the light-independent affect the light-dependent, or vice versa?
- What conflicting "needs" confront a plant living in a hot, dry environment? What is the frequent result in C3 plants? How is the problem avoided in C4 plants?
- What colors of the visible spectrum are absorbed by objects that are black? What about white objects? Does it really matter if you wear light colors when it is hot outside?
- **Connect to Chapter 46, Section 46.7 (The Carbon Cycle) and Chapter 48 (Human Impacts on the Biosphere**). Global warming has caused a rise in the world's oceans. How is this affecting world coastlines? How is this affecting the coastline closest to you? Can you envision some popular tourist locations that might be endangered by a rising ocean?
- What effect is global warming having on wildlife in the Arctic and in Antarctica?
- If global warming continues, will the areas where tropical diseases are prevalent expand northward?

Chapter 6 Homework Extension Questions

1. What role does water play in photosynthesis?
2. Explain the process used to capture light energy and convert it to stored chemical energy.
3. Summarize the role of the membrane in chemiosmosis in photosynthesis or cellular respiration.

Chapter 6 Possible Responses to *Critical Thinking* Questions

1. Let's look at the possibilities. The gain in weight of the tree *might* have been water simply absorbed through the roots and retained unchanged in the plant tissues, but *not* 164 pounds of it! Furthermore, we know that the water molecules are not going to combine *directly* with any other molecules in the tree. Of course, we now know much more about the reactions of photosynthesis than van Helmont did, so we can conclude that the gain in weight was from the synthesis of carbohydrate (specifically, cellulose) in the chloroplasts of the leaf cells due to the reaction of carbon dioxide from the air and water from the soil in the presence of light energy.
2. If you notice bubbles coming from an aquatic plant, the bubbles would contain oxygen, an end product of photosynthesis.
3. The first step in the Calvin cycle in C3 plants is the formation of two molecules of 3-phosphoglycerate (3-PGA) from CO_2 and ribulose bisphosphate (RuBP):

   $$RuBP + CO_2 \text{ -----> } 2\ 3\text{-PGA}$$

   So, the first compound that would incorporate the labeled C is 3-PGA.

For C4 plants, the first step in carbon fixation is the formation of oxaloacetate from CO_2 and phosphoenolpyruvate (PEP):

$$CO_2 + PEP \text{ ----> oxaloacetate}$$

So, the first compound that would incorporate the labeled C is oxaloacetate.

4. Since photosynthetic prokaryotes do not have membrane-bound organelles, including chloroplasts, the proteins necessary for the light-dependent reactions (including pigments such as chlorophyll) are located in the plasma membrane.

Chapter 6 Possible Responses to *Data Analysis Exercise* Questions

It is helpful when growing crops for use as biofuel to measure the net energy output, which is equal to the energy output minus the energy input.

1. The ethanol from one hectare of corn produced approximately 23 kcal $\times 10^6$. It took approximately 18 kcal $\times 10^6$ to grow the corn used to make ethanol.
2. The grass grown for synfuel had the highest ratio of energy output to energy input.
3. Corn would require the least amount of land to produce a given amount of biofuel energy.

CHAPTER 7: HOW CELLS RELEASE CHEMICAL ENERGY

Chapter 7 Suggestions for Presenting the Material

- Stress the elegance of the cycle that connects photosynthesis and aerobic respiration. The products of photosynthesis are the reactants of aerobic respiration, and the products of aerobic respiration are the reactants of photosynthesis. How has man offset the natural balance of this cycle?
- As you progress deeper into the pathway discussions, it is advisable to refer frequently to the overview graphic so that you keep the "big picture" in mind. You will need to indicate that fermentation begins with glucose and proceeds from pyruvate.
- Summing up the total energy yield is important.
- You can emphasize the roles of other foods (proteins and lipids) and their relationship to carbohydrates by following the arrows of Figure 7.12. This especially appeals to students interested in nutrition.
- Emphasize that plants do carry on aerobic respiration. Many students have the mistaken idea that plants *only* photosynthesize.
- Have students work on learning the reactants and products for each step of cellular respiration first. For example, glucose and two ATPs are the reactants for glycolysis; two pyruvates and four ATPs are the products. Once they have the reactants and products learned, the intermediate reactants are easier to learn and understand.
- Stress to students the importance of knowing the *processes* rather than memorizing all the various *reactions.*

Chapter 7 Classroom and Laboratory Enrichment

- **AP® Investigation #6: Cellular Respiration**
- Explain the purpose of lactic acid in muscle fatigue. Be sure to mention that once the oxygen debt has been repaid, the lactic acid breaks down into the harmless byproducts water and carbon dioxide, and physical activity can resume.
- Make a flowchart of the steps of cellular respiration in class, stressing the reactants and products of each step.
- Ask an exercise physiologist to talk to the class about the effect(s) of exercise on body metabolic rate.

- Discuss some animal examples of "red" and "white" types of muscles. A goose is an animal that seems to have almost unlimited endurance. If we cooked a goose for dinner, it would contain primarily dark meat. On the other hand, much of the tuna fish's meat is light in color, representing its ability to move in quick spurts of rapid movement. This connects to Chapter 35.
- Perform Activity 1 on the following website, http://www.accessexcellence.org/LC/SS/ferm_activity.php, which provides a fermentation experiment involving yeast.

Chapter 7 Classroom Discussion Ideas

- Certain diseases more common in the elderly have been tied to mitochondrial defects. Could mitochondrial degradation be a part of normal aging? How might you detect this in patients?
- In *Luft's syndrome*, the mitochondria are not producing sufficient amounts of ATP. What series of reactions could be most responsible for the deficiency? What are some other conditions that are thought to involve mitochondrial malfunction?
- Mitochondria are critical for normal metabolism. From which parent did each human being's original mitochondria come from at conception?
- Why are so many diseases attributed to defective mitochondria?
- Table wines, those that have *not* been fortified, have an alcoholic content of about 10–12%. What factors could limit the production of alcohol during fermentation? Is it self-limiting, or do the vintners have to stop it with some additive?
- Your text lists two types of fermentation: one leads to alcohol, the other to lactate. Which occurs in yeasts and why? Which pathway is reversible? What would be the consequences of nonreversible lactate formation in muscle cells?
- Yeast is added to a mixture of malt, hops, and water to brew beer—a product in which alcohol and carbon dioxide are desirable! Why is yeast added to bread dough?
- What happens to lactate produced during periods of intense muscle activity? Why is it important that the lactate is broken down quickly?
- What effect do large amounts of lactic acid have on muscle tissue?
- Think about which parts of a chicken are "dark meat." Considering the mitochondrial content of dark versus white meat, why does this make sense?
- What is "metabolic water"?
- Why is fermentation necessary under anaerobic conditions? That is, why does the cell convert pyruvate to some fermentation product when it does not result in any additional ATP production?

Chapter 7 Possible Responses to *Critical Thinking* Questions

1. Altitude sickness results from a lack of oxygen, but more specifically, it results from the deficiency of oxygen needed to "pull" the train of electrons along the electron transfer chain, which provides the energy necessary to produce ATP. This would explain why the symptoms of altitude sickness would mimic those of cyanide poisoning, namely the inability to produce enough ATP to power the muscle cells necessary for breathing and pumping blood.
2. Prokaryotes use the plasma membrane for electron transfer phosphorylation.
3. The journey of the bar-tailed godwit is indeed an amazing feat! This is accomplished by the majority of the flight muscles being of the dark meat variety. The darker color of the muscle is due to an abundance of myoglobin, which stores oxygen. This type of muscle is designed for prolonged activity due to its better endurance level.

Chapter 7 Possible Responses to *Data Analysis Exercise* Questions

The disease Tetralogy of Fallot (TF) is characterized by low oxygen levels and subsequent mitochondrial damage.

1. The abnormality in this study that was most strongly associated with TF was the number of mitochondria.
2. The percentage of TF patients that had mitochondria that were abnormal in size was 25%.
3. From the data provided, it appears that those patients that have both abnormal numbers and shapes of mitochondria are the most affected by the disease.

CONNECTING CHAPTERS 6 AND 7: PHOTOSYNTHESIS AND AEROBIC RESPIRATION

Chapters 6 and 7 Suggestions for Presenting the Material

- Emphasize the global connections between photosynthesis and cell respiration early in the unit, which will help students have a mental road map.
- Present the steps (light reactions, Calvin cycle, glycolysis, Krebs cycle, electron transport chain, and chemiosmosis) in manageable pieces.
- Use visual diagrams and analogies to illustrate the reactions involved in photosynthesis and cell respiration to assist students in understanding each step.
- Stress the interdependence among living things on Earth throughout the unit.
- Use concepts from previous units (structure and function of membranes, protein conformation) to make connections to current material (binding of oxygen to rubisco).
- This unit contains numerous details from which students can become disconnected, as they are not able to directly observe the reactions that occur. Create a unit-long investigation to assist students in staying connected to the content. Have students investigate how environmental conditions affect plant growth over time, rates of transpiration, or the effect of light of different wavelengths (using cellophane) on plant growth.

Chapters 6 and 7 Common Student Misconceptions

Many students get bogged down in the details of the reactions involved in cell respiration and photosynthesis while losing sight of the "big picture" and failing to make the vital connections between the two processes. Frequent connections to the ecological impact of the processes, as well as the influence of various factors such as temperature, light intensity, gas concentration, etc. on the rate of each process will assist in helping the students maintain perspective.

Students often do not make the connection that plants undergo both photosynthesis and cellular respiration. While plants release oxygen from photosynthesis, they also use some of that oxygen for aerobic respiration; therefore, plants have both mitochondria and chloroplasts.

Chapters 6 and 7 Classroom Discussion and Activities

Labs

- **Modify AP® Investigation #6 (Cellular Respiration)** to include photosynthesis by using a photosynthetic organism (like plants) in various light conditions. Use a system that can measure either CO_2 or O_2 in real time.
- Students can study the interdependence of living things on each other and the energy and nutrients in their environment by designing controlled investigation using a mini-habitat study with a small aquatic ecosystem (snail, water plants, etc.) or terrarium.
- See critical thinking question #1, Chapter 6, (Starr/Taggart; *Biology: The Unity and Diversity of Life*, 15th edition) to discuss the research completed by Jan Bapista van Helmont regarding how trees obtain the materials to increase their size. This discussion will reinforce the concepts of the exchange of matter presented in the previous unit.
- Investigate how temperature affects the rate of cricket chirping using recorded chirping.
- Investigate how exercise affects the amount of CO_2 produced using a diluted solution of bromothymol blue.
- Have students work in small groups to model the energy coupling that occurs with photosynthesis and cell respiration. You may want to assign a specific set of reactions to each group. Encourage visual representations of the connection, such as role plays.
- Have students contrast the differences between chemiosmosis in cell respiration and photosynthesis. One method for presenting the comparison would be to give students a copy of an image of the process in each location and create a T-chart from what they identify as similarities and differences in the diagram. Peer editing or review can be used to allow communication between groups to broaden the activity. This can be done as a previewing activity before discussing chemiosmosis or as a culminating/summarizing activity after the discussion.

Chapters 6 and 7 Lesson Closure

- Global warming is a popular topic in the mainstream media, and students likely have misconceptions and questions. After studying the patterns of energy flow in cell respiration and photosynthesis, students are prepared for a more in-depth investigation on the effects of climate change. This is an excellent opportunity for students to practice data analysis, make logical inferences, and develop their own evidence-based conclusions using global climate data, carbon dioxide graphs, and research-based sources. This culminating activity can take many forms depending on the class and time available. Independent research projects, portfolios, class debates, small or whole group discussions, or small group presentations can be used to investigate global warming.

Case Studies

The following case studies are from National Center for Case Study in Teaching Science (http://sciencecases.lib.buffalo.edu/cs/collection/):

- The Case of the Dividing Cell
- My Dog Is Broken
- You Are Not the Mother of Your Children
- The Case of Ruth James
- Osmosis Is Serious Business

- A Rigorous Investigation
    - Clarifies misconceptions about cellular respiration
- The Mystery of Seven Deaths
    - Discussion of the function of cellular respiration and the electron transport chain
- "Why Is Patrick Paralyzed?"
    - A rare genetic disease in which an enzyme is deficient

AP® Practice Essays

1. Discuss the structure and function of the components of the cell membrane.
    a. Pick three methods of transporting substances across the membrane.
    b. Explain what happens to a red blood cell when placed in a hypertonic and hypotonic environment.
    c. Certain structures between neighboring cells within the same tissue help to facilitate material exchange. Describe this using one example structure each from a plant and an animal.
2. Water is vital to the survival of all living things. Cells must constantly balance its movement across the membrane.
    a. Explain how osmosis occurs in cells.
    b. Describe THREE different coping strategies used by groups of organisms that allow them to manage the water in their environment.
    c. Describe how an organism might exploit osmotic forces.
3. Enzymes are vital to the metabolism of all cells.
    a. Describe how their form is related to their function and what conditions could alter their form, rendering them useless.
    b. Describe three mechanisms that enzymes use to lower the activation energy of a reaction.
    c. Explain how allosteric enzymes can be involved in feedback inhibition.
4. Aerobic respiration and anaerobic fermentation can be used by cells to produce ATP.
    a. Distinguish between the two types of fermentation.
    b. Describe how these two pathways are different.
    c. Describe the major advantages of each.
5. The structure of a chloroplast allows it to capture light energy and transform it into chemical energy.
    a. Describe how the structure of the chloroplast allows it to carry out these reactions.
    b. Describe the noncyclic and cyclic pathways of ATP formation.
    c. Explain what conditions would cause the chloroplast to use the cyclic pathway of ATP formation.
    d. Explain how the light-independent reactions produce carbohydrates.

Lesson Outline: Unit 2a: Inheritance—Genetics: Gene Expression and Biotechnology

Correlates with part of Unit 2 in the 15th edition book, Chapters 8–10 and 15

- **AP® Biology Big Idea 2:** Biological systems utilize free energy and molecular building blocks to grow, reproduce, and maintain dynamic homeostasis.
- **AP® Biology Big Idea 3:** Living systems store, retrieve, transmit, and respond to information essential to life processes.
- **AP® Biology Big Idea 4:** Biological systems interact, and these systems and their interactions possess complex properties.

Brief chapter summaries

Chapter 8 ("DNA Structure and Function") gives historical relevance to the discovery of DNA as the molecule of inheritance, as well as how the DNA in a chromosome is replicated into two sister chromatids during S phase of mitosis (DNA duplication).

Chapter 9 ("From DNA to Protein") describes the importance of genes in their role for coding for proteins through the processes of transcription and translation. Chapters 8 and 9 do a good job of describing the biological basis of Big Idea 3 as it applies to genetics.

Chapter 10 ("Control of Gene Expression") puts transcription and translation into the context of real-world responses by giving examples that answer questions, such as *when* do different types of genes get expressed and how much? How do cells respond to certain environmental stimuli? This chapter also has relevant ties to the use of developmental patterns to help determine evolutionary relationships (Section 18.5) and human development (Chapter 42). Both of these topics are a part of the College Board's AP® Biology curriculum.

Chapter 15 ("Studying and Manipulating Genomes") introduces students to how we can use what we know about how genes function to learn more about genomes, especially as applied to human endeavors, such as studying and treating genetic diseases and creating genetically modified organisms.

CHAPTER 8: DNA STRUCTURE AND FUNCTION

Chapter 8 Objectives

- Name the three components of nucleotides and the four nitrogen bases found in DNA. Indicate which are purines and which are pyrimidines.
- Describe the Watson and Crick model. Show how this model accounts for precise genetic replication.
- Define the sequence of a new strand of DNA that is replicated from a given strand of DNA.

Enduring Understanding 3.A: Heritable information provides continuity of life.

Essential Knowledge 3.A.1: DNA, and in some cases RNA, is the primary source of heritable information.

Chapter 8 Warm Up Questions

1. What is the molecular basis of inheritance?
2. When during the cell cycle does DNA replication occur?
3. How can a biologist account for such variation within our species?
4. How can a biologist account for variation from species to species?

Chapter 8 Lesson Opener

Speedy: (10–15 minutes) Begin by discussing the difference between asexual and sexual reproduction. How does a population acquire genetic variation? Discuss the slipper limpet process of reproduction and discuss what the evolutionary advantages are. Discuss the process of parthenogenesis in salamanders—how does this limit genetic diversity? Show the short film clip from *The Mating Game on Trials of Life*.

Extensive: (40–50 minutes) Show short film clip about Dolly (www.youtube.com) and discuss how the cloning process takes place. Have a DNA model to show the helical structure. Open with a short film clip (10 minutes) on DNA replication. Discuss why DNA must make an exact copy of itself.

Abstract article: "The Blue People of Troublesome Creek," by Cathy Trost, *Science*, November 1982. Have the students read the article and discuss mode of inheritance and the concept of the founder effect (Link to Evolution: Chapter 17).

Show a 20-minute film clip from *Gattaca*. Discuss the significance of the film title and how would it be possible to determine the genetic makeup of one's children? What are some moral and ethical issues?

Have students extract DNA from their cheek cells. Lab DNA Extraction: http://www.seplessons.org/node/222.

Chapter 8 Suggestions for Presenting the Material

- This chapter amplifies the information on nucleic acids presented at the close of Chapter 3. Depending on the amount of information you presented in your lectures at that time, some of this chapter could be repetitious. Be sure they remember that all nucleotides have a sugar, a nitrogenous base, and one or more phosphates. Use diagrams and/or an identification game.
- For best success in presenting this chapter, use diagrams, models, and figures when discussing the structure and replication of DNA.
- This is a good time to remind students that scientists need to be aware of the social impact of their work. At various times, the U.S. government has placed limits on stem cell research. This chapter provides the opportunity to have a discussion about the rights a government should have over such areas of research, the ethics of stem cell research, and/or the ethics of using stem cells to treat diseases.
- Ask students to think about the benefits and drawbacks of DNA as a genetic material.
- Search online for computer software that allows the user to build and/or manipulate DNA models on the screen.

Chapter 8 Classroom and Laboratory Enrichment

- Ask students to work in teams of two. Give each student a set of labeled paper shapes representing the sugars, phosphate groups, and each of the four bases present in DNA. Ask each student to construct a short segment of a DNA strand while his or her partner builds the complementary strand of the DNA double helix. Then ask students to demonstrate the semiconservative replication of DNA.
- Use a video to demonstrate the semiconservative nature of DNA replication. There is also an excellent website by the Howard Hughes Medical Institute Foundation (HHMI) that includes some of the best interactive clips of DNA replication, transcription, and translation I have ever used. This information can be used here and in Chapter 12: http://www.dnai.org.
- DNA is described as a "double helix" or "twisted ladder." An inexpensive device that can show this structure very well is a plastic parakeet ladder that is flexible enough to be twisted from "ladder" configuration to "helix."

- If you search for DNA replication on the Internet, you will find many great animations. Here are two of the many that I have found: http://www.johnkyrk.com/DNAreplication.html and http://www.umass.edu/molvis/bme3d/materials/jtat_080510/exploringdna/contents/contents.htm (interactive).
- Another website of interest featuring tutorials along with problem sets based on DNA's structure is http://www.biology.arizona.edu/biochemistry/activities/DNA/DNA.html.
- Prepare a chronological listing of the dates, people, and significant contributions to the discovery of the structure of DNA.

Chapter 8 Classroom Discussion Ideas

- Why don't the *different species* of single-celled and multicellular organisms have *different nucleic acids* for coding hereditary information? Why do they all use DNA as the hereditary material?
- What are the benefits touted by those who advocate cloning? What are the objections raised by those who oppose cloning?
- What are some reasons DNA is double-stranded instead of single-stranded?
- What are some advantages of semiconservative replication?
- Of the experiments done before the structure of DNA was known, which ones demonstrated that nucleic acid was the carrier of heredity?
- Rosalind Franklin collected data critical to the elucidation of DNA structure. However, she is hardly mentioned in textbook accounts. Locate a biography of her and speculate on why she is lesser known than her collaborators.
- Why should the term *DNA relative* replace the more popular term *blood relative* when referring to human kinship?
- Which of the following is a more likely source of altered DNA sequences?
  a. A new copy has an error made during replication from a correct original.
  b. A new copy has an error faithfully copied from an incorrect original.
- A *casual* reading of any one of a number of biology texts would imply that Fred Griffith was a pioneer in DNA research. Is this an accurate assessment?

Chapter 8 Homework Extension Questions

- What would the shape of a DNA molecule be like if purines paired with purines and pyrimidines paired with pyrimidines?
- Why do eukaryotic cells "need" *histones*, *nucleosomes*, and *looped regions*?
- How did the Hershey and Chase experiment settle the question of which molecule—DNA or protein—carries heredity?

Chapter 8 Possible Responses to *Critical Thinking* Questions

1. The replication would look like this (dark lines = ^{15}N; light lines = ^{14}N)

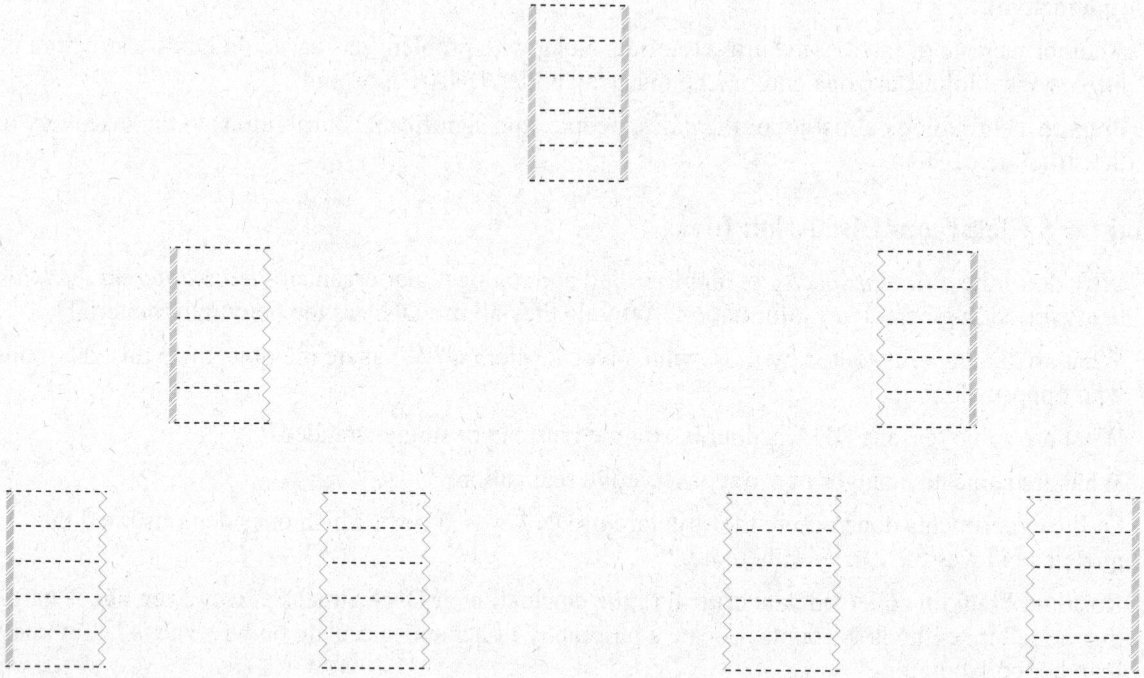

2. As submitted in the movie *Jurassic Park*, something always goes wrong when you mess with nature. Therefore, the con to cloning a woolly mammoth, first of all, would be the question, "Why do we wish to do this?" If the animal went extinct, it is for a good reason through the processes of natural selection. What benefit is there in bringing back an animal that has been selected out of existence? I would also add another question to the con side—how much money would this cost, and can this be justified in light of other ongoing medical research needs? The pro side might argue that in science we are always learning, and to see a slice of the history of Earth preserved and then made whole again in the woolly mammoth would provide valuable details about the conditions necessary for its survival in prehistoric times. The process would be a way of gathering data, which science is always doing. It would be fascinating to study a living mammoth: for example, to be able to examine its physiological makeup and compare this to its descendants, the African and Asian elephants.
3. Normally, when exposed to UV radiation, the thymine dimers in the DNA molecule are identified and excised, then replaced with undamaged nucleotides produced during protein synthesis of interphase. The corrected copy is passed on during DNA replication. Individuals with xeroderma pigmentosum (XP) have no enzymes available to repair the thymine dimers. With defects in the repair processes, the resulting mutations lead to defects in the corresponding gene products, which lead to all the symptoms of XP.

Chapter 8 Possible Responses to *Data Analysis Exercise* Questions

The Hershey-Chase experiment was designed to detect which portion of a virus entered infected bacteria. The protein coat of the virus was labeled with sulfur and the DNA was labeled with phosphorus.

1. Before blending, the percentage of ^{35}S was approximately 5% and the percentage of ^{32}P was around 16%.
2. After 4 minutes, there was 80% of ^{35}S and 30% of ^{32}P.

3. The researchers knew that the radioisotopes in the fluid did not come from broken bacteria, because the number of bacteria remained fairly constant.
4. The extracellular concentration of ^{35}S increases the most with blending. These results indicate that viruses inject DNA into the bacteria, because the percentage of extracellular ^{32}P stays low in extracellular fluid. This shows that the phosphorus, which labels DNA, is largely inside the infected bacteria.

CHAPTER 9: FROM DNA TO PROTEIN

Chapter 9 Objectives

- State three differences between DNA and RNA.
- Name the types of RNA and indicate where each is synthesized and where each is active.
- Discuss the function of each of the following in protein synthesis:
  a. DNA
  b. RNA
  c. rRNA
  d. mRNA
  e. tRNA
  f. ribosomes
  g. amino acids
- Describe the process of transcription and translation.
- Define *mutation* and give four examples of how mutations can occur.
- Describe the structure of a virus and the mechanisms of reproduction (lytic and lysogenic cycles).

Enduring Understanding 3.B: Expression of genetic information involves cellular and molecular mechanism.

Essential Knowledge 3.B.1: Gene regulation results in differential gene expression, leading to cell specialization.

Chapter 9 Warm Up Questions

1. How might alterations in DNA structure be harmful to a species? How might such alterations be beneficial? What type of genetic change is most important for evolution?
2. Ask students to compare and contrast: transcription and translation; codons and anticodons; and rRNA, mRNA, and tRNA.
3. Why is transcription necessary? Why don't cells use their DNA as a direct model for protein synthesis?

Chapter 9 Lesson Opener

There are many excellent YouTube videos that depict the process of transcription and translation. After first articulating why transcription and translation occur, describe the following analogy in which the process is compared to the construction of a building.

a. DNA "sealed" in the nucleus	a. Master blueprints that remain in the architect's office
b. mRNA that leaves nucleus to go to ribosome	b. Blueprint copies taken to the job site
c. Ribosomes and rRNA	c. The construction site
d. Enzymes	d. Construction workers
e. tRNA carrying amino acids	e. Trucks carrying materials
f. Amino acids	f. Building materials
g. Protein	g. Building

Have groups of students come up with their own, different analogy.

Chapter 9 Suggestions for Presenting the Material

- The subject of protein synthesis is a difficult one even when presented on an introductory level. Begin by very briefly summarizing the process: DNA $\xrightarrow{\text{transcription}}$ RNA $\xrightarrow{\text{translation}}$ protein. You may also want to review protein structure (Chapter 3).
- Students find it hard to understand and identify the components of DNA, so begin this section of the text by making sure they have a clear picture of deoxyribose (the five-carbon sugar in DNA), phosphate groups, and the four nitrogen bases. Briefly show diagrams of the molecular structure of these three major players, and then introduce the term *nucleotide*.
- Students should be able to achieve a good understanding of protein synthesis if they begin by visualizing it as two major steps, transcription and translation, rather than getting lost in complex details. The events of protein synthesis can be effectively presented with visual aids such as diagrams, animations, and models. Students need to have some kind of mental picture in order to understand what happens during the making of a protein.
- Stress the role that mutations play in producing "new" genes and "new" proteins.
- To show how insertions and deletions might affect a reading frame, write this sentence on the board: THE RED FOX RAN FAR. To show the effects of insertions, insert a letter after "T" and have your class read the sentence (try A). It should read TAH ERE DFO XRA NFA R. Try deleting a letter, such as the letter "D," showing the effects of a deletion. The new sentence should read THE REF OXR ANF AR.

Chapter 9 Classroom and Laboratory Enrichment

- Use models, figures, and animations to show protein synthesis.
- Ask students to think about the benefits and drawbacks of DNA as genetic material.
- Have students come up with a skit to act out the processes of transcription and translation.

- The following items may help your students remember the difference between "transcription" and "translation."
  a. *Transcription* involves the transfer of information from one form to another *in the same* language, for example, an office memo in shorthand transcribed into typed copy but both in English; likewise, a section of genetic code in DNA is copied to RNA (both nucleic acids).
  b. *Translation* is the transfer of information in *one language* to *another language*, for example, a story in French to English; likewise, genetic code in RNA is transferred to amino acids (nucleic acid to protein).
- Discuss the various causes and agents of mutations to help students become aware of how easy it is to be exposed to mutagens. You can refer back to the topic of cancer at this time, as many mutagenic agents are known to also be carcinogenic. A common cancer that occurs when an individual is overexposed to UV radiation is skin cancer.

Chapter 9 Classroom Discussion Ideas

- Describe the three stages of translation.
- In what ways are the instructions encoded in DNA sometimes altered?
- In most species, mutation is usually not considered a driving evolutionary force. Why?
- In what ways does RNA differ from DNA?
- This chapter refers to the participants and process involved in protein synthesis as if they have been *seen* doing their work; have they? How then do we know all this information is accurate?
- How can you explain the occurrence of birth defects (caused by altered genes) in children and grandchildren of the victims of the atomic bombs that destroyed Hiroshima and Nagasaki, Japan, when the victims themselves were only mildly affected?
- How is the movement of a portion of DNA in the process called *transposition* different from *translocation*?
- Describe experiments performed by Khorana, Nirenberg, Ochoa, Holley, and others to decipher the genetic code.
- Discover why repeated applications of a single drug or pesticide can result in resistance among bacterial strains and species of insects. Why does this pose a problem? What steps can be taken to avoid resistant strains of pathogenic bacteria and disease-carrying insects? Link to artificial selection in the Evolution unit.
- Prepare a chart that graphically depicts the series of errors (in DNA, mRNA, tRNA, and amino acids) that lead to the production of the abnormal hemoglobin in sickle-cell anemia.
- The progress in molecular biology has proceeded from deciphering genetic codes to the construction of manmade genes by machine. Report on the construction and use applications of such devices.
- In addition to its importance in protecting development of fetuses, folic acid has been claimed to prevent many different types of cancers, as well as neurodegenerative disorders such as Alzheimer's disease and Parkinson's disease. Folate is involved in photorepair, where thymine dimers are corrected. Research folic acid and find out its role in cellular processes and why a deficiency in folic acid can lead to disorders and disease.
- What are some common substances that act as mutagens? How are some of these mutagens known to cause cancer? Are there substances that will block the effects of mutagens?
- Do some research to find out more about ricin. By what mechanism does ricin inactivate a ribosome?
- Find out more about the castor oil plant that produces ricin. Why might a plant need to store a toxin as lethal as ricin in its tissues and seeds? (Hint: You can discuss the "evolutionary arms race" here and expand your discussion when you discuss evolutionary adaptations in Unit 3.)

Chapter 9 Homework Extension Questions

1. Which of the RNAs are "reusable"?
2. Why do you think DNA has *introns*, which are transcribed but removed before translation begins?
3. How does a *chromosomal variation* (Chapter 13) differ from a *gene mutation*?
4. If all DNA is made of the same basic building units (sugar, phosphate, and nitrogenous bases), how can DNA differ in, say, a human and a bacterium?

Chapter 9 Possible Responses to *Critical Thinking* Questions

1. Antisense drugs interact with their intended target based on information in the genetic code. Overproduction or abnormal production of proteins is implicated in cancer and many diseases. Antisense drugs will stop the translation of these proteins by preventing the mRNA coding for them from reaching the ribosome. They do this by binding (hybridizing) to the target mRNA, which is degraded by enzymes and thus not translated.
2. Nucleic acid sequences are traditionally written in a 5′→3′ direction, so the anticodon sequence is 5′-GCG-3′. This anticodon would bind to 5′-CGC-3′, which according to Figure 9.7 encodes arginine. If the C of the anticodon were changed to a G, the tRNA would still carry the arginine amino acid; but the anticodon now binds to a 5′-CCC-3′ codon, which encodes proline. In this case, the mutant tRNA would be competing with the normal proline tRNA, and in some cases an arginine would be inserted instead of proline.
3. You could not have fewer than the three-base codons that exist. If you had a two-base codon, the maximum number of different codons would be 16 ($4 \times 4 = 16$). The three-base codons allow for 64 different codons ($4 \times 4 \times 4 = 64$). This is more than the 20 amino acids and allows for redundancy, meaning that multiple codons encode some amino acids. If you look closely at Figure 9.7, you will notice that, in most cases, the redundancy is in the third base of the codon. Thus, if a mutation occurs in the third base of a codon, it will not change the amino acid sequence of the protein.
4. The following is the given RNA segment (hyphens have been added for easier reading):

   5′-GGU-UUC-UUC-AAG-AGA-3′

   Consult Figure 9.7 (genetic code) to see which amino acid matches each codon to produce the finished amino acid sequence of the polypeptide:

   glycine-phenylalanine-phenylalanine-lysine-arginine
5. If translation begins at the second base, the codons are as follows:

   5′-GUU-UCU-UCA-AGA-GA-3′

   The resulting protein is valine-serine-serine-arginine.
6. The "stem-loop" structure would interfere with the RNA polymerase because it is working in a linear fashion as it "ratchets" down the DNA strand. The polymerase is pulled away from its working position because of an incorrect fit as the hydrogen bonds between the AU base pairs are disrupted.

Chapter 9 Possible Responses to *Data Analysis Exercise* Questions

1. The skin cancer cells show the greatest response to the engineered RIP. Notice that the cell survival decreases as the concentration of the RIP increases in concentration.
2. At 10^{-7} grams/liter, all the cells survive.
3. The breast cancer cells. At this concentration, the skin cancer cells are greatly reduced in survival, and the liver and prostate cells are slightly reduced in survival.
4. The data points linked by straight lines are cells that do not have a dose-dependent response to the concentration of RIP.

CHAPTER 10: CONTROL OF GENE EXPRESSION

Chapter 10 Objectives

- Know the various ways that gene activity (transcription and translation) are turned on (activated) and off (inactivated).
- Know the difference between promoters and enhancers.
- Describe the controls before transcription, during processing and translation, and after translation.
- Describe how cells differentiate by selective gene expression.
- Know what X-chromosome inactivation is and how it relates to dosage compensation.
- Explain the ABC model and how it relates to flower formation.
- Describe how homeotic genes control body-plan formation.
- Explain the regulation of the lac operon.
- Describe how humans develop lactose intolerance.
- Explain what epigenetic marks are and how they can be passed to offspring.
- Understand how cancer may result when regulatory genes are altered.

Enduring Understanding 4.A: Interactions within biological systems lead to complex properties.

Essential Knowledge 4.A.3: Interactions between external stimuli and regulated gene expression result in the specialization of cells, tissues, and organs.

Chapter 10 Warm Up Questions

1. Why are calico cats always female?
2. How might alterations in DNA structure be harmful to a species?
3. What is the role of gene control in causing cancer?
4. What is the difference between a negative gene control and a positive gene control?

Chapter 10 Lesson Opener

Begin by asking the students what would happen if all of the genes in a human muscle cell were being transcribed and translated at the same time. Get them to the point where they'll realize that all human cells *have* all the genes needed for every function and, therefore, it wouldn't act as a muscle cell anymore. Chapter 10 gives a few examples of how cells "decide" which genes get turned on and when in both eukaryotes and prokaryotes.

Chapter 10 Suggestions for Presenting the Material

- This chapter builds on information that students learned in previous chapters about gene structure and function. Terms such as DNA, mRNA, transcription, and translation must be familiar before beginning this chapter.
- Emphasize to the students that the control of gene expression is an extremely complex subject area, one which is best approached by first studying some fairly simple and well-understood examples in prokaryotes.
- Give students opportunities to learn and use new words such as promoter, operator, and operon. Gene control among eukaryotes will be easier to understand if students view it as a series of levels.

- Students have a difficult time relating to and understanding gene regulation because it is so foreign to their life experiences. Using analogies might help them relate to the information. Here is an analogy you can try involving a car:

  Repressors are the parking brakes; they inhibit any movement on transcription until they are released.

  The binding of inducers, like lactose, is analogous to releasing the parking brake. Now the car is able to move forward—but RNA polymerase must bind to the promoter.

Chapter 10 Classroom and Laboratory Enrichment

- Use visual aids such as overhead slides to illustrate the lactose operon (Figure 10.10).
- Use models to show induction and repression of gene expression in the operon.
- Have the students compare and contrast prokaryotic and eukaryotic gene control.
- If you can arrange a demonstration of chromosome puffs in fruit flies by geneticists, do so; it is a good visual aid for seeing transcription in action. If the timing is inconvenient for students to actually see the process, make a visual recording of the procedure to show later.
- Modify Figure 10.10 (lactose operon) by obscuring the labels. Ask students to identify each item on the figure.
- Prepare a summary table that lists the following:

Type of Control	Specific Example	Found in:		
		Prokaryote	Eukaryote	Both
a. Transcriptional				
b. Transcript processing				
c. Translational				
d. Post-translational				

Chapter 10 Classroom Discussion Ideas

- What are some of the possible environmental agents that could trigger cancer?
- Why do you think the disease cancer and deaths from cancer receive so much attention by Americans?
- Evaluate this comment, "Cancer is the disease of those people who live too well."
- What are the benefits of learning you have a gene that may lead to cancer? What are the disadvantages of receiving such knowledge?
- What is the likelihood that there will be a breakthrough cure for cancer in your lifetime?
- How might alterations in DNA structure be harmful to a species? How might such alterations be beneficial? What type of genetic change is most important for evolution?
- What is the role of gene control in causing cancer? How are some viruses known to be linked to cancer?
- Scientists know much more about controls over gene expression among prokaryotes than among eukaryotes. What are some reasons why research in this area is more difficult among eukaryotic species than it is among prokaryotic species?
- Do you think cancer-causing genes could someday be repaired?
- There is much more DNA in eukaryotic cells than scientists call necessary. Why do *you* think it is there?
- Why do eukaryotic cells "need" *histones*, *nucleosomes*, and *looped domains*?
- What would be the hypothetical effect on the lactose operon of a modified lactose molecule? Do you think it would still bind to the repressor?
- What is the "economic" advantage to a prokaryotic cell of possessing inducible enzymes?

- Distinguish between negative gene control and positive gene control.
- Why are so many of the homeotic genes similar among animal groups? What group would have homeotic genes the most similar to those of humans?
- Learn more about current research efforts attempting to uncover the mysteries of development and differentiation.
- Describe the operon and its function.
- Learn more about oncogenes and cancer.
- Discover more about the discovery and diagnostic uses of the Barr body in female mammalian cells. What is the current research saying about X inactivation?
- What are some common substances that act as mutagens? How are some of these mutagens known to cause cancer? Are there substances that will block the effects of mutagens?
- Learn more about tumor cell lines used to study cancerous cell growth *in vitro* in the laboratory.
- Locate the original article by Francois Jacob and Jacques Monod proposing the lac operon. How have the details changed?
- Likewise, see if you can locate the original research publications of Murray Barr and Mary Lyon. Notice the dates of these publications. Were they before or after the publication of DNA structure by Watson and Crick in 1953?
- Research current work of evolutionary geneticists with homeotic or Hox genes in humans.

Chapter 10 Homework Extension Questions

1. Why are some genes expressed and some not expressed?
2. Do the same gene controls function in bacterial cells and humans? Why or why not? Explain in a paragraph why a woman is considered a "mosaic" for the expression of genes on the X chromosome.
3. Construct a chart showing the four major types of control (transcriptional, transcript processing, translational, and post-translational)—give a specific example of each and indicate whether the example is found in eukaryotes, prokaryotes, or both.
4. Why is the SRY gene important to males?

Chapter 10 Possible Responses to *Critical Thinking* Questions

1. In adult humans, many genes that were required for proper development are no longer needed and are permanently turned off (not expressed). For example, hemoglobin, the protein that carries oxygen in the blood, is a large protein that consists of four smaller proteins. Since a developing fetus lives in a reduced oxygen environment, fetal hemoglobin must bind oxygen at a high affinity. Fetal hemoglobin contains two copies of alpha globin protein and two copies of gamma globin protein. Thus, the genes that encode alpha and gamma globin are expressed during early development. At birth, a newborn enters an environment that is higher in oxygen, and different globin genes are expressed to match the changing environmental conditions. Beginning at birth and continuing through adulthood, hemoglobin is made of two alpha globin proteins and two beta globin proteins. Thus, at birth, the gamma globin gene is turned off and the beta globin gene is turned on. In this example, gene expression changes to match a change in environmental conditions.
   Genes are also differentially expressed because different tissues perform different functions and thus require different proteins to be made. For example, muscle cells require the expression of genes that encode proteins necessary for muscle contraction. Pancreatic cells require the expression of genes that encode proteins necessary for making insulin. Since these are specialized tissues, muscle cells do not express genes that make insulin and pancreatic cells do not express the genes for muscle contraction. It should be noted that there are genes that are expressed in all tissues. These genes encode proteins that are required for basic cellular function (e.g., DNA replication, transcription, and translation).

2. Individuals that are starving express genes that encode proteins that will increase fat production. This is a survival mechanism that allows the individual to produce as much fat as possible under famine conditions. These fat production genes may be regulated epigenetically. In a starving parent, the fat production genes are turned on because they have less DNA methylation than an individual living in normal conditions. The offspring of the starving parents could inherit the low methylation state, and even if the child is raised in a normal environment, the child's fat production genes are overactive because of the inherited methylation state.
3. No. Three types of gene controls are (1) control by histone organization of the DNA, (2) transcriptional control by regulatory proteins in operons, and (3) control of the transport of mRNA out of the nucleus. Those not found in prokaryotes are (1) because the single, circular DNA of prokaryotes is not organized around histone proteins and (3) because there is no nucleus in prokaryotes. Of course, (2) is very typical of prokaryotes; the same mechanism is not found in eukaryotes.
4. According to the data, individuals with *BRCA1* are diagnosed with cancer at an earlier age and have the highest percentage of death. While there are far fewer individuals who have the *BRCA2* mutation, the percentage that dies is much lower. Thus, *BRCA1* is the more dangerous breast cancer gene.

   To determine the effectiveness of the preventative surgeries, one would have to evaluate how many of those who died from breast cancer had the surgeries and how many did not. In addition, it would be helpful to know the age that the individuals died and their age of diagnosis. This information might reveal that preventative surgery increases survival time even if it does not prevent death.

Chapter 10 Possible Responses to *Data Analysis Exercise* Questions

1. A girl whose grandmother experienced a famine at age 2 is one and one-half times more likely to die early.
2. The data show that, in general, when a grandmother experienced a famine, her grandchildren are at an increased risk of mortality. If a grandmother was well-fed, her grandchildren are at decreased risk of mortality. However, this trend is reversed around age 9. If a maternal grandmother was well-fed around age 9, her grandchildren have an increased risk of early mortality. If a maternal grandmother experienced a famine at age 9, her grandchildren have a reduced risk of early mortality.
3. These data suggest that the sex chromosomes are involved in the epigenetic changes being inherited. It could be that the epigenetic changes are located on the sex chromosomes. Alternatively, the sex chromosomes could simply be influencing epigenetic changes on the non sex chromosomes (autosomes).

CHAPTER 15: STUDYING AND MANIPULATING GENOMES

Chapter 15 Objectives

1. Know how DNA can be cleaved, spliced, cloned, and sequenced.
2. Understand what plasmids are and how they may be used to insert new genes into recombinant DNA molecules.
3. Know the role of restriction enzymes and DNA ligase in forming recombinant DNA.
4. Understand what a cDNA library is and how it is created and the role RNA transcriptase plays in this process.
5. Explain how gene libraries are produced, and describe the role of probes, primers, and nucleic acid hybridization in this procedure.
6. Explain what PCR is, and describe its use in replicating genes.
7. Describe automated DNA sequencing, and explain how it reveals the sequence of PCR-amplified DNA.

8. Understand what DNA profiling is, and how it can be used for medical and forensic purposes.
9. Understand how the Human Genome Project has paved the way for the field of genomics.
10. Explain how genomics furthered the development of human gene therapy.
11. Describe how SNP chips are used to detect genomic differences between individuals.
12. Explain what genetic engineering is, and how it is used to manufacture drugs and clean up the environment.
13. Describe how genetic engineering has helped the agricultural community.
14. Explain how transgenic animals are used in medical research.
15. Name the ethical issues concerning transgenics.
16. Describe the safety concerns and precautions taken in DNA research.
17. Describe eugenics and eugenic engineering and the pros and cons of human gene therapy.
18. Explain the types of genetic experiments that nature has been performing for billions of years.
19. Understand how one organism can produce the products of another.
20. Be aware of several limits and possibilities for future research in genetic engineering.

Chapter 15 Warm Up Questions

1. What does "genetic engineering" mean?
2. Why can the green fluorescent protein (GFP) gene found in jellies still "work" in bacteria?

Chapter 15 Lesson Opener

One class activity you might wish to consider before or during discussing Section 15.10 is showing clips from the movie *Gattaca* (1997). Parents meet with their geneticist to select *which* egg and sperm they wish to unite through fertilization to become their child. In this meeting, they are able to select for sex, intelligence, diseases, physical appearance, and abilities, if they so choose—in fact, the exact date and cause of natural death for every human can be determined from a single drop of blood. Such a clip can be a good discussion launcher into future possibilities in eugenic engineering.

Chapter 15 Suggestions for Presenting the Material

- Help students see the relevance of this subject by telling them about some of the products (such as insulin) that are produced as a result of genetic engineering.
- Begin by reminding students that genetic recombination occurs naturally in all organisms during meiosis. Emphasize that even though examples of genetic research in bacteria may seem obscure and of little relationship to more complex eukaryotic genomes, such experimentation yields results of great value to humans.
- Ask questions to ensure that students are indeed knowledgeable about the use of plasmids and restriction enzymes. Visuals are very helpful with this rather abstract and unseen procedure.
- Genetic engineering has the potential to truly capture the students' interest. You may wish to devote extra class time to this subject in order to allow students the opportunity to become involved and invested in it. More than likely there will be issues about genetic engineering they will be voting on in the future.

Chapter 15 Classroom and Laboratory Enrichment

- **AP® Investigation 8: Biotechnology: Bacterial Transformation**
- **AP® Investigation 9: Biotechnology: Restriction Enzyme Analysis of DNA**
- Prepare a summary table of the genome modification methods listed in this chapter. Include the following information:
  - Natural versus manmade
  - Examples of organisms
  - Usefulness
- Ask two groups of students to prepare brief arguments *for* and *against* the continuation of genetic engineering research and development.
- Arrange for students to visit a colleague's laboratory where PCR is regularly performed to get an idea of the technology through a demonstration.

Chapter 15 Classroom Discussion Ideas

- What are some characteristics of an organism ideally suited for research in genetic engineering?
- Discuss the benefits of genetic engineering versus potential risks.
- What (if any) might be the objections to growing and distributing the "golden rice" in third-world countries experiencing food shortages?
- Do you think that new genomes resulting from genetic engineering should be patented? Who should receive monetary benefits from such discoveries—the research scientists performing the work or their academic institutions?
- What advantages would insulin produced by genetic engineering have over preparations from animal sources in the treatment of human diabetes mellitus?
- Is genetic engineering a new concept in nature or just the human application of a natural mechanism already in operation? Explain with examples.
- Is it dangerous to eat foods that have genes from another organism?
- Describe problems that have resulted from the standard prophylactic use of antibiotics among farm animals such as poultry, pigs, and cattle.
- Describe the safeguards currently followed in labs doing work in genetic engineering.
- Have students research the rapidly growing field of *proteomics*.

Chapter 15 Homework Extension Questions

1. Where and when does genetic recombination naturally occur?
2. Why are restriction enzymes useful tools for genetic engineering? How is DNA ligase used in genetic engineering?
3. Discuss the growing problem of antibiotic resistance among the different species of bacteria responsible for causing diseases such as gonorrhea, typhoid, and meningitis.

Chapter 15 Possible Responses to *Critical Thinking* Questions

1. DNA bacteriophages are viruses that infect bacterial cells by injecting their DNA in host bacterial cells. Once the phage DNA enters the bacteria, it can take over the host cell's machinery to produce more bacteriophages inside the bacterial cell. Once a large number of phage progeny are made inside the host cell, the cell bursts open and the phage progeny are free to infect other bacterial cells. Restriction enzymes protect the bacteria from bacteriophages by chopping up the phage DNA when it

enters the host cell. The host cell's DNA is protected because bacterial DNA is methylated. Most restriction enzymes will not cut at its recognition sequence if that sequence is methylated.

2. If a talking chimp could be engineered, it is likely that the debate over using animals in research would intensify. While there are many protocols in place to ensure that research animals are not subjected to unnecessary pain and suffering, in many ways we are only able to speculate if the research is causing pain or suffering. In addition, the definition of unnecessary pain and suffering is not always clear. Currently, there are committees that evaluate the use of animals in research and ensure that the research is conducted in a way that minimizes the risks to the animal subjects and maximizes the benefits of the research. If the chimps could vocally communicate their pain and suffering in our language, the protocols for animal research would likely be reevaluated since the amount of pain and suffering to research animals would be more clearly understood. The issue is where do we draw the line? Where do the benefits of animal research (and there are many) outweigh the pain and suffering? If an animal can't communicate in our language, is it still ok to perform the research even after the talking chimps have communicated their pain and suffering? Or do we eliminate research on talking chimps but continue performing research on animals who can't talk?
3. Viruses are very complex and in many ways not well understood. It is known that most viruses naturally mutate very quickly. There is a possibility that modern influenza viruses could become more deadly on their own, as was demonstrated with the H1N1 strain in 2009. The benefits gained in the research performed on deadly viruses far outweigh the risks. There is plenty of data and many viruses available for terrorists to use for horrific purposes. Our best defense against bioterrorism attacks is to better understand how viruses function.

Chapter 15 Possible Responses to *Data Analysis Exercise* Questions

1. A little over 5 days.
2. The modified mice; they learned the location in 4 days.
3. The modified mice; they learned the location in about 2.5 days.
4. Since the mice are using visual cues to help them remember where the platform was located in previous trials, and since the modified mice found the platform faster than the unmodified mice, the data indicate that the modified mice had the greatest improvement in memory.

Chapters 8–10 and 15 Unit Closure

- Give a brief overview of the Human Genome Project and its director Dr. Francis Collins. How did this project help to identify the location of certain detrimental genes? How will this knowledge impact scientists to be able to help people with these detrimental genes in the future? https://www.genome.gov/10001772/all-about-the--human-genome-project-hgp/
- Discuss the role of gene control in causing cancer. How can alterations in DNA be either harmful or helpful to a species? This discussion of mutations leads into the next section of evolution and natural selection, as mutations are the source of evolution.
- BLAST activity: Connecting DNA to Disease: https://teacherknowledge.wikispaces.com/Quist+-+Connecting+DNA+to+Disease+Using+BLAST

Common Student Misconceptions

- Students find it difficult to understand and identify the molecular components of DNA and why replication is continuous on one strand but discontinuous on the other strand.
- Students struggle to keep separated the ideas of DNA replication, transcription, and translation and when they occur in a cell. A flowchart showing the central dogma with bullets of when each process is used is very helpful for students to connect back to throughout the unit.

- Students have the misconception that all mutations are "bad." The teacher needs to relate mutations to genetic diversity and evolution. Examples of mutations that can be beneficial are important to cement their understanding.
- Students have difficulty in comprehending gene regulation in prokaryotic and eukaryotic cells. Acting out an operon through role-play can clarify the process.
- Students think that dominant alleles are "better." Discuss specific examples of harmful dominant alleles such as polydactyly, achondroplasia, Huntington's, and progeria.

Concept Reinforcement Labs

- Wartski/Nicholson, Teacher's Manual for Advanced Placement Course in Biology; 1997/Duke TIP. www.tip.duke.edu: Lab: Chromatin Isolation Using Onions
- Wartski/Nicholson, Teacher's Manual for Advanced Placement Course in Biology; 1997/Duke TIP. www.tip.duke.edu: Lab: Amylase Regulation in Prokaryotes
- Recovering the Romanov's—Investigative CD lab using forensics. Wards Science.
- Teaching Chi Square Analysis Using M&M's. Tim Ligget, Conestoga HS, Berwyn, Pa.
- Have each student research a trait and prepare a family pedigree.
- Have students construct actual karyotypes. Wards Science.
- DNA from the beginning: Animated primer of 75 experiments that made modern genetics. www.dnaftb.org
- Great source for DNA, genetics, protein synthesis—hands-on visual labs: http://learngenetics.utah.edu/.
- Wards: Simulated Disease Transmission Lab—how a virus can be spread through body fluids.
- A Bewildering Tale: A Molecular Crime Drama; Charlotte McBee/Jennifer Dye, NSTA 1999.
- Teaching Molecular Genetics Using Paper DNA Sequences; NABT 1998; Scott MacClintic: Scott_MacClintic@loomes.org.
- The Blackout Syndrome: Access Excellence Science Mystery sponsored by Genetech, Inc. http://www.accessexcellence.org/tbs

Films

- Films for Humanities and Science—Protein Synthesis
  a. Protein: The Staff of Life
  b. DNA: The Molecule of Heredity
  c. DNA: Replication and the Repeating Formula
  d. RNA Synthesis: The Genetic Messenger
  e. Transfer RNA: The Genetic Messenger
  f. Ribosomal RNA: The Protein Maker
- Biomedia Associates Films: DNA Replication and Mitosis
- Biomedia Associates Films: The Genetic Code and Its Translation: Gene Regulation in Prokaryotes

Novels

- *The Genesis Code*, John Case. Thriller concerning the cloning of Jesus using DNA from religious relics.
- *Survival of the Sickest*, Dr. Sharon Moalem. Common disorders are connected through evolution, inheritance, and environmental pressures.

- *Genome*, Matt Ridley.
- *Darwin's Ghost*, Steve Jones. Introduction contains explanation of HIV's evolution and how its genetic processes enhance its ability to survive

Case Studies

National Center for Case Study Teaching in Science. http://sciencecases.lib.buffalo.edu/cs/

- PKU Carriers
- A Sickeningly Sweet Baby Boy
- Bad Blood
- Can a Genetic Disease Be Cured?
- Giving Birth to Someone Else's Children
- In Sickness and Health
- Agony & Ecstasy
- Cloning Man's Best Friend
- Bringing Back Baby Jason

AP® Practice Essays

1. Compare and contrast the gene regulation process in the lac operon in bacterial cells to a few gene regulation processes in eukaryotes.
2. Every somatic cell in your body is an exact copy of the cell that came before it. Your DNA has been copied by trillions of cells over and over again.
   (a) Describe the structure of a DNA nucleotide.
   (b) Explain how DNA is replicated in a eukaryotic cell.
   (c) Describe how DNA is sorted into two new nuclei during mitosis and meiosis.
3. The flow of genetic information begins with DNA and ends with expressed traits. Describe the process of producing proteins beginning with messenger RNA leaving the nucleus and ending with a polypeptide disengaged from the ribosome.

Lesson Outline: Unit 2b: Inheritance—Cell Reproduction and Mendelian Genetics

Correlates with 15th edition book, Chapters 11–14

- **AP® Biology Big Idea 2:** Biological systems utilize free energy and molecular building blocks to grow, reproduce, and maintain dynamic homeostasis.
- **AP® Biology Big Idea 3:** Living systems store, retrieve, transmit, and respond to information essential to life processes.
- **AP® Biology Big Idea 4:** Biological systems interact, and these systems and their interactions possess complex properties.

Brief chapter summaries

Chapter 11 ("How Cells Reproduce") describes how one cell can copy and distribute all of the chromosomes into two daughter nuclei, and then how the cytoplasm divides.

Chapter 12 ("Meiosis and Sexual Reproduction") describes how haploid gametes such as sperm and eggs can form from diploid germ cells in preparation for sexual reproduction.

Chapter 13 ("Observing Patterns of Inheritance") is often a student favorite. Students learn various patterns of inheritance such as simple dominant/recessive traits and are taught how to predict offspring genotype and phenotype ratios using Punnett squares.

Chapter 14 ("Chromosomes and Human Inheritance") introduces students to the modes of inheritance of various human traits, as well as determining those modes of inheritance via pedigree analysis. Chromosomal aberrations are also mentioned, such as trisomy 21.

Links to Chapter 20 ("Viruses and Bacteria") gives an overview of how viruses and bacteria live and reproduce. It is a good idea to compare and contrast viral, prokaryotic, and eukaryotic processes whenever possible.

Objectives

- Explain the difference between asexual and sexual reproduction and why sexual reproduction leads to genetic variation.
- Define and use the terms *allele*, *segregation*, F_1, and F_2.
- Distinguish between the following pairs of terms: monohybrid/dihybrid, homozygous/heterozygous, and dominant/recessive.
- Explain how a test cross is performed and why it is a useful genetic tool.
- Use a Punnett square and probabilities to perform monohybrid and dihybrid crosses.
- Explain how incomplete dominance differs from complete and codominance.
- Give the characteristic phenotypic ratio of the F_2 in a dihybrid cross in which the two genes are independent.
- State Mendel's first and second laws and relate them to the chromosomal theory of inheritance.
- Distinguish between sex chromosomes and autosomes.

- Define *linkage*, and explain how crossing over creates genetic diversity.
- Show how crossing-over frequencies are calculated and how they can be used to make chromosome map.
- Explain how translocation, deletion, duplication, inversion, and nondisjunction alter chromosomes.
- Discuss how certain genes are carried on the X or Y chromosome and so are linked.
- Identify whether a certain condition is inherited as autosomal recessive or dominant.

CHAPTER 11: HOW CELLS REPRODUCE

Chapter 11 Objectives

- Explain the process of the cell cycle.
- Explain the importance of checkpoints in the cell cycle.
- Explain how the structure and quantity of DNA changes throughout the cell cycle.
- Distinguish between sexual and asexual reproduction.
- Explain the need for cell division.
- Relate the development of cancer cells to the cell cycle.
- Contrast cytokinesis in plant and animal cells.
- Identify which events in meiosis create genetic diversity.

Enduring Understanding 3.A: Heritable information provides for continuity of life.

Essential Knowledge 3.A.2: In eukaryotes, heritable information is passed to the next generation via processes that include the cell cycle and mitosis or meiosis plus fertilization.

Chapter 11 Warm Up Questions

1. How do cells reproduce?
2. Do all cells reproduce?
3. Predict the steps a cell would need to take to prepare for dividing.

Chapter 11 Lesson Opener

As a student project, encourage the construction of a model of the chromosome. Perhaps some wire and plastic spools could be the building materials. Research one of the human autosomes and illustrate or model the chromosome by showing the names and locations of the genes that have been identified.

Chapter 11 Suggestions for Presenting the Material

- This is the first of two chapters concerning cell reproduction. The present one explains mitosis—division in which the number of chromosomes remains the same in the identical daughter cells. The next chapter explains a more complicated type of cell division in which the number of chromosomes is reduced during the production of cells destined to become gametes—meiosis. Students need to have a solid understanding of mitosis before learning meiosis.
- Emphasize the cell cycle, stressing that cells are not dividing all of the time. Students should be aware of the fact that the steps of cell division are part of a continuum. Our separation of the process into four stages is an artificial one, and it may be hard to say where one stage ends and the next begins when looking at a dividing cell. You may compare it to the showing of a "game tape" that athletes watch to see the errors committed during the big game.

- Students often have trouble following the number of chromosomes throughout the stages of mitosis. To help them, remind them that each chromosome has one centromere, and that it is not until centromeres divide in anaphase that *sister chromatids* are considered *chromosomes*. Make certain that students understand where and when mitosis occurs in any organism.
- Another approach to explaining the *number* of chromosomes before and after mitosis is to abandon all mention of chromosomes and simply keep track of the number of DNA molecules present at each stage. It would look like this for a human somatic cell: G_1 (46), S (92), G_2 (92), prophase (92), metaphase (92), anaphase (46 moving to each pole), and telophase (46 each cell).
- Emphasize to the student that the "secret" to preserving the chromosome number in the two daughter cells of mitosis and cutting the chromosome number in half in the four daughter cells of meiosis (next chapter) is the same—DNA duplication during interphase!
- A good way for students to count chromosomes in the diagrams is to count the centromeres of the chromosomes.
- Ask students what would happen if cytokinesis did not occur.

Chapter 11 Classroom and Laboratory Enrichment

- **AP® Investigation #7: Part 1: Modeling Mitosis, Part 2: Effects of Environment on Mitosis, and Part 3: Loss of Cell Cycle Control in Cancer.** This lab also connects to the AP® Science Practice #2: Using Math.
- Show a video of time-lapse photography of cells undergoing actual cell division. There is a great video available at http://www.bozemanscience.com/cell-division/.
- Ask students (working individually or in small groups) to use chromosome-simulation kits (available from biological supply houses) to demonstrate chromosome replication during the stages of mitosis. If kits are unavailable, make your own chromosomes using pipe cleaners and following the instructions provided in this online activity: http://www.indiana.edu/~ensiweb/lessons/gen.mm.html.
- View ready-made squashes of mitotic material, such as onion root tip, under a microscope or project slides of such material. Ask students to estimate the length of each mitotic phase after counting the number of cells in each phase in several fields of view.
- Using photos of the stages of mitosis, project each stage (not in correct sequence) and ask students to identify each phase and give one or two major events happening in each phase.
- Perhaps the following analogy can help students visualize chromosomes during mitosis:
  two chromatids = two nearly matched socks
  one centromere = one clothespin
  spindle fiber = clothesline

Chapter 11 Classroom Discussion Ideas

- Loosely speaking, the process of one cell becoming two cells is often referred to as mitosis, but to be completely accurate, what does mitosis *specifically* refer to?
- Why do cells undergoing mitosis require one set of divisions but cells undergoing meiosis need two sets of divisions?
- Why are the HeLa cells considered to be immortal? Are there any considerations when it comes to researching cancer on the basis of HeLa cells (e.g., contamination)?
- What have been some of the ethical issues surrounding the use of HeLa cells? See http://www.smithsonianmag.com/science-nature/henrietta-lacks-immortal-cells-6421299/ for discussion. Also see the book "The Immortal Life of Henrietta Lacks" by Rebecca Skloot available at http://rebeccaskloot.com/wp-content/uploads/2011/03/HenriettaLacks_RGG.pdf.

- Using a generation time of 20 minutes, calculate the size of a bacterial population that has arisen from a single bacterium growing under optimum conditions for 8 hours (e.g., *Salmonella* in a bowl of unrefrigerated potato salad at a picnic on a warm summer day).
- Many of the drugs used in chemotherapy cause loss of hair in the individual being treated. Ask students if they can figure out why such drugs affect hair growth. What are some new developments in preventing hair loss for cancer patients?
- Biologists used to believe that interphase was a "resting period" during the life cycle of the cell. Why did this appear to be so? What do we now consider interphase to be?
- Ask students how cell division in plant cells differs from that in animal cells.
- How can there be 46 chromosomes in a human cell at metaphase and also 46 chromosomes after the centromere splits in anaphase? Hint: Focus on the name change of chromatids to daughter chromosomes.
- Why do students think chromosome number varies across species and what might this tell us about the evolutionary history of organisms? How might mitosis and changes in errors potentially give rise to new species (e.g., in plants)?
- What is there about the composition of an animal cell versus a plant cell that necessitates different methods of cleavage?
- Explore diseases (such as cancer) that involve cell growth gone wrong. How do such diseases affect the mechanism of cell division? What drugs are used to halt runaway cell growth? How do these drugs work, and what are their side effects?
- Why do some cells of the human body (e.g., epithelial cells) continue to divide, yet other cells (e.g., nerve cells) lose their ability to replicate once they are mature? Describe some of the latest research efforts to induce cell division in nerve cells.
- Colchicine is a chemical used to treat dividing plant cells to ensure that chromosomes of cells undergoing mitosis will be visible. How does colchicine achieve this effect? What is the natural source of colchicine?

Chapter 11 Possible Responses to *Critical Thinking* Questions

1. No. Cells can remain in the G_1 state of the cell cycle. Instead of preparing to divide and produce more cells, the cell simply performs its specialized function. For example, a pancreatic cell may receive instructions to produce insulin. Ultimately, the cell will die and need to be replaced by other cells that have been instructed to undergo cell division.
2. The mystery cell is an animal cell. It is evident due to the presence of a contractile ring. Plant cells do not have the contractile ring mechanism but instead deposit vesicles of cellulose to form the cell plate. See Figure 11.7 for further details.
3. Radiation has the greatest effect on cells with high rates of mitosis, such as those in hair follicles, the lining of the gut, and cancer cells. Radiation kills cancer cells since they are constantly dividing.
4. One way to measure the amount of DNA in the cell is to count the number of DNA molecules at each point in the cell cycle. It would look like this for a human somatic cell: G_1 (46), S (92), G_2 (92), prophase (92), metaphase (92), anaphase (46 moving to each pole), and telophase (46 each cell). Thus, you would see a change in the amount of DNA per cell during the S and M phases of the cell cycle.

Chapter 11 Possible Responses to *Data Analysis Exercise* Questions

1. 82
2. 36. A normal human cell has 46 chromosomes.
3. Since there is no Y chromosome, this cell is from a female.

Chapter 11 Homework Extension Questions

1. Summarize the life cycle of a cell.
2. How is the cell cycle controlled to maintain accuracy?
3. Explain how mitosis conserves the chromosomal number and meiosis reduces it.

CHAPTER 12: MEIOSIS AND SEXUAL REPRODUCTION

Chapter 12 Objectives

- Examine the advantages of sexual reproduction over asexual reproduction.
- Examine the characteristics of an allele.
- Outline the stages of meiosis using a diagram.
- Analyze the processes of meiosis I and meiosis II by using diagrams.
- Assess the influence of crossing over and chromosomal segmentation on the inheritance of variable genetic traits.
- Outline the process of gamete formation in plants and animals using diagrams.
- Examine how mitosis and meiosis are similar.

Enduring Understanding 3.A: Heritable information provides continuity of life.

Essential Knowledge 3.A.3: The chromosomal basis of inheritance provides an understanding of the pattern of passage (transmission) of genes from parent to offspring.

Enduring Understanding 3.C: The processing of genetic information is imperfect and is a source of genetic variation.

Essential Knowledge 3.C.2: Biological systems have multiple processes that increase genetic variation.

Essential Knowledge 3.C.3: Viral replication results in genetic variation, and viral infection can introduce genetic variation into the host.

- *Note:* These two Essential Knowledges connect nicely with parts of Chapter 20.

Chapter 12 Warm Up Questions

1. Why do some children look a lot like one or both of their biological parents, whereas others don't look very similar to their biological parent(s)?
2. How does the process of meiosis apply to Mendel's Law of Segregation?
3. Did Mendel know anything about DNA when he was developing his ideas about inheritance?
4. How does sex generate variation? How do bacteria reproduce?
5. How do bacteria, which asexually reproduce, increase genetic variation? (link to Chapter 20)
6. How can viruses change the genetic makeup of an organism? (link to Chapter 20)

Chapter 12 Lesson Opener

Using pop beads or any other chromosome simulation, give students a picture of a diploid "cell" with the four chromosomes (two homologous pairs) in G_1 of S phase, and another picture of four haploid "cells" post-meiosis. Ask them to come up with what the chromosomes might "look like" in G_2 (after DNA replication), and then each stage of meiosis. Ask them to draw what they look like at each stage in order to get the chromosomes to look like the "after" picture. After they have bumbled through a set of guesses, ask them to compare their pictures with the textbook and do it again correctly.

Chapter 12 Suggestions for Presenting the Material

- This is a crucial subject area for beginning biology students. Students must have a good understanding of meiosis to comprehend the workings of inheritance, explained in subsequent chapters.
- Before beginning this chapter, ask questions to make sure students are well grounded in the events and purpose of mitosis. Because it was covered in the previous chapter, students should still recall the terms used to describe the parts of the cell involved in cell division.
- Learning and understanding the terminology are critical for students to comprehend the differences between meiosis and mitosis. Make sure to review the difference between germ cells and somatic cells, germ cells and gametes, haploid and diploid, homologous and homolog chromosomes, autosomes and sex chromosomes, etc.
- The events of meiosis can be confusing. Emphasize that meiosis makes it possible for organisms to undergo sexual reproduction. Remind students that the benefit of sexual reproduction is genetic diversity. This will help them understand why a process as complex as meiosis has evolved.
- Students often find it hard to understand when and how the chromosome number changes during meiosis, so be sure they understand that the two chromatids of one chromosome are each considered a chromosome in their own right *after* the centromere splits during meiotic anaphase II.
- Meiosis will be easier to grasp if students can become thoroughly acquainted with a typical animal life cycle (use the human life cycle as a familiar example). Before finishing with this chapter, be sure to question the students about the events of meiosis and its consequences to the organism.
- Be sure to show and explain the excellent comparison of mitosis and meiosis in Figure 12.11.

Chapter 12 Classroom and Laboratory Enrichment

- **AP® Investigation #7: Part 4: Modeling Meiosis and Part 5: Meiosis and Crossing Over in *Sordaria***
- Demonstrate the phases of meiosis, using a series of progressive slides or a video. A good resource is "Meiosis" from the Molecular and Cellular Biology Learning Center at http://vcell.ndsu.nodak.edu/animations/meiosis/movie-flash.htm.
- Show students a slide of a karyotype of a normal man or woman to introduce the concept of homologous pairs.
- Compare a human karyotype to that of another organism.
- Ask students (working individually or in small groups) to use chromosome-simulation kits (available from biological supply houses) to demonstrate chromosome replication and reduction of chromosome number during the stages of meiosis. If kits are unavailable, make your own chromosomes using a pop-it bead (or bead and cord) for each chromatid and a magnet for each centromere. A cheap and easy alternative is using colored craft sticks (available in all craft stores). The color denotes maternal or paternal origin, and you can create various sizes of chromosome by cutting the sticks.

- Illustrate crossing over by using lengths of different colored string. Snip and tie the ends to create the products of a crossover (see Figure 12.6). Another option is to use colored modeling clay. After rolling out strips of two different colors (maternal versus paternal), form homologous pairs and then select the chromatids to participate in the crossing-over event. Tear the dough and reassemble it to show the exchange. Students can also do this in a lab setting.
- Point out the subtle differences between the generalized life cycles of plants and animals.
- Show students a video depicting the union of gametes and subsequent cleavage; it will provide visual presentation of what mitosis and meiosis accomplish in a living cell. See http://www.hhmi.org/biointeractive/human-embryonic-development.
- Show the effects of crossing over through pop-bead kits. Use round stickers for the six different pairs of alleles and place them along two sets of chromosomes (stick each one to one of the pop beads). Have students initiate a random crossover while finishing modeling meiosis. Have students write down their allele combination in their gametes. Have them do the crossover differently each time, while still writing their allele combination for each of their gametes. Ask the class for the different allele combinations they saw in their gametes. Remind students that crossing over does not have to occur, but in humans it typically occurs twice along each chromosome.

Chapter 12 Classroom Discussion Ideas

- Why do cells undergoing mitosis require one set of divisions, but cells undergoing meiosis need two sets of divisions?
- If bacteria reproduce asexually and give rise to clones, how do antibiotic-resistant bacteria develop? Why do bacteria not die out when their environment changes? (This is an excellent question that will come up again in the evolution unit.)
- Many species reproduce asexually, some species can reproduce asexually and sexually, and humans and the majority of other species reproduce only sexually. What are the advantages and disadvantages associated with different modes of reproduction?
- What would be an advantage of parthenogenesis? What are some unexpected species that are capable of parthenogenesis (e.g., sharks, reptiles, and fish)?
- Division of the cell cytoplasm is equal during spermatogenesis but unequal during oogenesis. Can you think of at least one reason why?
- Why are mules and hinnies considered infertile?
- An old-fashioned name for meiosis is "reduction division." Why?
- How does crossing over introduce new genetic variation?
- One of the meiotic series is very much like mitosis. Is it meiosis I or II?
- Does the reduction in chromosome number occur in meiosis I or II? Hint: To conveniently count the number of chromosomes (whether doubled as chromatids or newly formed daughter chromosomes), simply count the number of centromeres (or portions thereof) present in any particular stage.
- When do the processes of human spermatogenesis and oogenesis begin? Are they the same in males and females?
- What is the derivation of the prefix "chrom-" as used in describing the threadlike bearers of hereditary instructions?
- Ask students to compare an animal life cycle to a complete land-plant life cycle. How is the life cycle of a human different from that of a plant?
- The generalized life cycle of complex land plants is often described as "alternation of generations." Describe the meaning of this phrase.

Chapter 12 Possible Responses to *Critical Thinking* Questions

1. If meiosis fails in generation 1, the gametes would have 26 chromosomes (instead of 13). If these abnormal gametes combine, the individuals in generation 2 would contain 52 chromosomes. If meiosis fails in generation 2, those individuals would have gametes that contain 52 chromosomes. If these gametes combine, the individuals in generation 3 would contain 104 chromosomes.
2. Sexual reproduction, specifically meiosis, sometimes (but not always) allows for greater genetic diversity. A more salient point is that sexual reproduction, through recombination, can break apart genetic associations in a population. This is favorable when environmental conditions are rapidly changing. Also, recombination can result in different genes from different individuals to be brought together, generating new combinations that may improve the response to selective pressures in the environment. However, it is not necessarily correct that sexual reproduction by default generates greater variation.
3. Sexually reproducing snails accumulate mutations faster, and this is expected to be even greater in polluted systems. As a result, sexually reproducing snails will likely be far less successful in polluted fresh water systems.
4. In the following figures, the chromosomes are shown during the various phases of meiosis. For simplicity, crossing over and independent assortment are not indicated in the figures.

For 2N = 4:

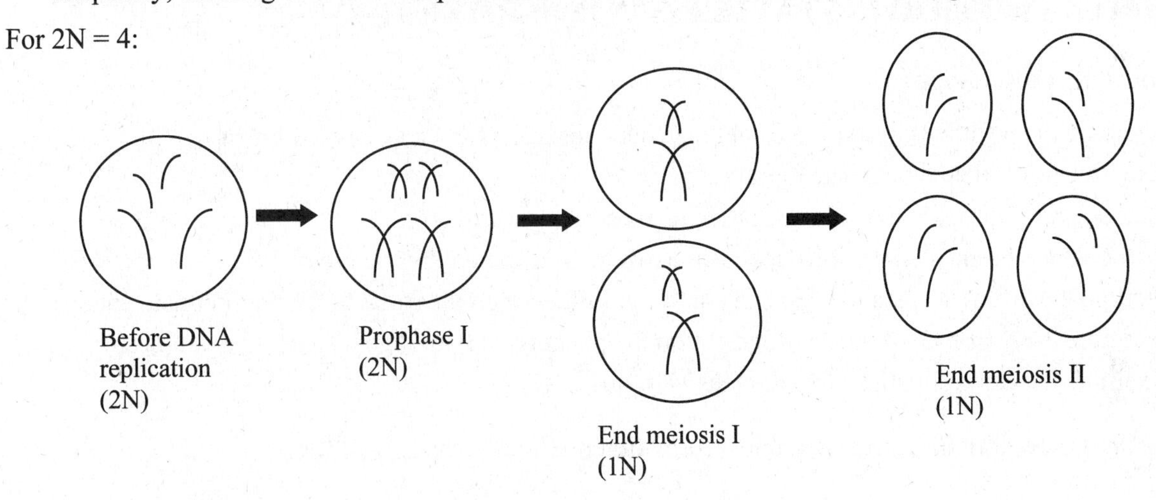

For 2N = 2 + 1 (there is an extra copy of one chromosome):

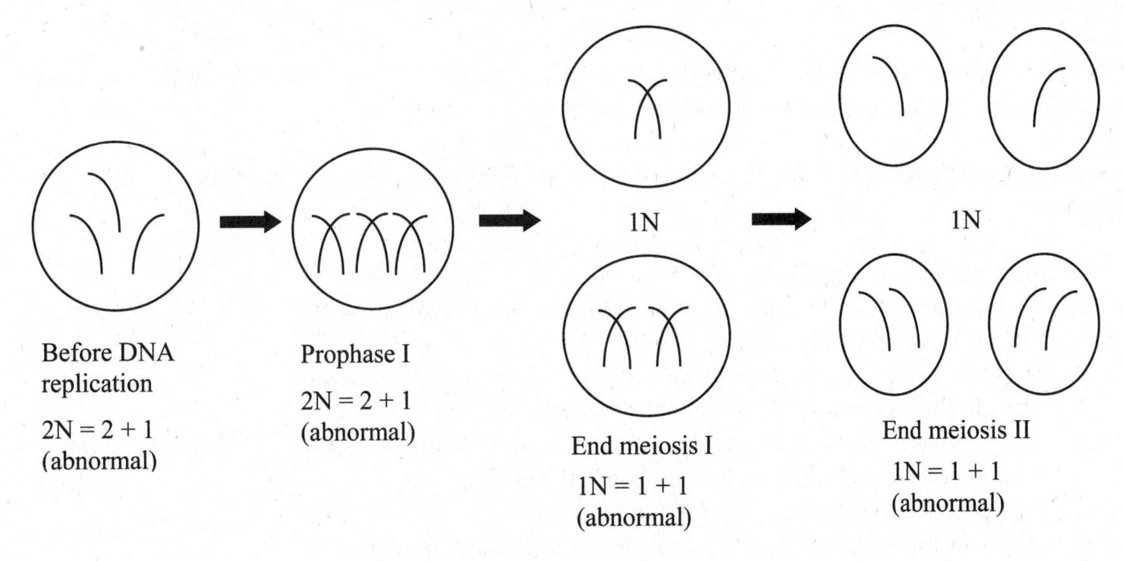

Chapter 12 Possible Responses to *Data Analysis Activities* Questions

1. 1.8%
2. The mice exposed to damaged cages with damaged bottles.
3. During metaphase, all the chromosomes should be aligned on the metaphase plate as shown in Figure 12.11A. In Figure 12.11B, the chromosomes are not aligned properly on the metaphase plate. In Figure 12.11C, there is one chromosome that has been excluded and is not on the metaphase plate. In Figure 12.11D, there are two groups of chromosomes instead of one.

Chapter 12 Homework Extension Questions

1. Pictorially represent random assortment and crossing over.
2. Why does crossing over occur in prophase of *meiosis* and not in *mitosis*?
3. What is the difference between conjugation, transformation, and transduction?
4. What would happen in the formation of the zygote if meiosis did not halve the chromosome number?
5. Compare the ways in which genetic variation is introduced into the population of sexually reproducing species, those that can reproduce via parthenogenesis, as well as bacteria and viruses.

CHAPTER 13: OBSERVING PATTERNS IN INHERITED TRAITS

Chapter 13 Objectives

- Examine the negative consequences of the genetic disorder known as cystic fibrosis.
- Examine how alleles contribute to traits.
- Demonstrate Mendel's law of segregation using a monohybrid cross.
- Demonstrate Mendel's law of independent assortment using a dihybrid cross.
- Outline the different ways in which an allele can influence an inherited trait using Punnett squares.
- Examine the influence of the environment on the phenotype of an organism using examples.
- Examine the characteristics of continuous variation.

Enduring Understanding 3.A: Heritable information provides continuity of life.

Essential Knowledge 3.A.4: The inheritance pattern of many traits cannot be explained by simple Mendelian genetics.

Enduring Understanding 4.C: Naturally occurring diversity among and between components within biological systems affects interactions with the environment.

Essential Knowledge 4.C.2: Environmental factors influence the expression of the genotype in an organism.

Chapter 13 Warm Up Questions

1. Blood types used to be presented as evidence for or against paternity lawsuits. Why might this be inconclusive evidence?
2. Why are there more boys who are born with color-blindness than girls?
3. Are all alleles completely dominant or recessive?
4. Given what you know about inheritance, account for the phenotypic differences within the human species for traits such as height and weight?
5. If skin color is heritable, explain tanning in the sun.
6. Why are transplanted organs sometimes rejected by the recipient?

Chapter 13 Lesson Opener

Distribute PTC tasting paper to your students, and calculate the number of tasters and nontasters in the classroom. You can also use different physical traits such as tongue rolling or earlobe attachment. Connect the differences in phenotypes to the differences in the coding regions of the genes. It's a good idea to mention that in the next unit they will learn how to use math to calculate the estimated numbers of people with particular allele combinations.

Chapter 13 Suggestions for Presenting the Material

- Students are usually naturally curious and interested in genetics. Start first with the simple examples of Mendel's monohybrid and dihybrid crosses before fielding questions on human traits, such as height or eye color. Emphasize the remarkable nature of Mendel's work; remind the students that he knew nothing of chromosomes and their behavior, and that the term *gene* did not exist until several years after his death.
- Use Mendel's experiments and his conclusions as real-life examples of the scientific method at work. Ask questions to make sure students understand monohybrid and dihybrid crosses and testcrosses.
- Emphasize genetic terms and the figures that make use of these terms. Use Figure 13.5 to ensure that students can visualize homologous chromosomes, gene loci, alleles, and gene pairs.
- Students should be able to relate the events of meiosis to the concepts of segregation and independent assortment; if their understanding of meiosis is weak, they will have trouble doing this.
- Many students come into college biology classes with the misconception that dominant alleles are "better" or more common than the recessive alleles. This misconception also makes it difficult for these students to understand non-Mendelian patterns of inheritance discussed at the end of Chapter 13. Stress to your students that dominant does not mean the most abundant or most advantageous trait. Give some examples of complete dominance with alleles that are harmful, undesirable, or lethal, such as polydactyly (extra digits), achondroplasia (dwarfism), and Huntington's disease (lethal degenerative neurological disorder).
- Beginning with this chapter, students will be quick to ask questions about human traits, many of which are governed by mechanisms more complex than those postulated by Mendel. Answer questions in this area during (or after) the discussion of variations on Mendel's themes, presented in the second half of this chapter.
- Remind students that the statistical probabilities obtained from Punnett squares are not absolute. There will be some variance, but large sample sizes reduce the likelihood.

Chapter 13 Classroom and Laboratory Enrichment

- Ask groups of students to conduct coin tosses. Demonstrate the importance of large sample size by having the students vary the number of tosses before calculating variation from expected ratios.
- Expand the biographical sketch of Mendel, including his education and practice as a clergyman. Enliven your presentation with as many slides of photos as you can find.
- Hand out a partially completed pedigree, and show students how to assign squares and circles for their family. Then ask them to select a trait and complete the pedigree after surveying the family members for presence/absence of the trait.
- Select a portion of the class to reenact the photo in Figure 13.17A. If the quantity of students chosen does not provide a bell-shaped curve, use this as an illustration of how the greater number of trials/subjects/experiments tends to increase probability.
- Show this TEDEd video on genetics: http://ed.ted.com/lessons/how-mendel-s-pea-plants-helped-us-understand-genetics-hortensia-jimenez-diaz.

- Have students do a couple of quick practice Punnett squares (Tt × tt). Students should try to explain the given probabilities and describe possible phenotypes.
- There are many interesting websites that give possible phenotypes of offspring from parents that have different eye color; this shows students the complexity of polygenic inheritance.
- Demonstrate cross-fertilization by bringing two large ornamental flowers to class and performing the cross in front of the students.
- An activity to help demonstrate how genotype governs phenotype can be a nice break from lecture. Using the example of fictitious genes may seem oversimplified and silly to you, but students will enjoy the break. To show incomplete dominance: The "L" gene codes for laughter, the "L" allele codes an enzyme producing very loud laughter, and the "l" allele codes for an enzyme that produces no laughter.

  Without explaining what you are demonstrating, hand two students papers with *L* and have them come to the front of the class. Hand two other students papers with *l* and have them come to the front, too. Point out that each student represents one chromosome that is carrying an allele for the *L* gene. Pair the chromosomes (students) by grouping like alleles (*LL* and *ll*). Then have the students demonstrate their phenotypes: *LL* students laugh loudly and *ll* students remain silent. Now, illustrate segregation by having the alleles (student groups) separate; show fertilization when an *L* and an *l* come together. Within each new group, have the *L* students laugh while the *l* students remain silent. Point out that the volume of laughing in the *Ll* individual is *about half* what it was in the *LL* individual. This is how a trait showing incomplete dominance works.

  You can extend this to include epistasis: CCHH, CCHh, CcHH, CcHh—no hand clapping despite the "C" allele, because the "H" allele is present and inhibits the "C" allele from expressing (student can hold the clappers' hands); CChh, Cchh—hand clapping, no inhibitor is present; ccHh, ccHH, cchh—no hand clapping because no "C" allele is present.

Chapter 13 Classroom Discussion Ideas

- Describe the behavior of one trait with regard to its inheritance in a particular cross, then ask students to identify the genetic mechanism at work (simple dominance, recessive inheritance, incomplete dominance, codominance, epistasis, pleiotropy, and polygenic inheritance).
- Cystic fibrosis has been targeted for gene therapy. Discuss the pros and cons of using gene therapy to help cystic fibrosis patients.
- Discuss the pros and cons of requiring individuals to be genetically screened for disorders such as cystic fibrosis. Ask the students to think about how their opinion on having children might change if they knew they were carriers.
- List some human traits that you would guess are governed by a single gene.
- Give several reasons why Mendel's pea plants were a good choice for an experimental organism in genetics. Give an example of an organism that would be a poor choice for genetic research and explain your choice.
- Conduct a quick review in class of the various inheritable traits. Ask the students what kind of inheritance pattern the trait is an example of.
- Describe several different crosses using organisms such as Mendel's pea plants. Then ask students to calculate phenotypic and genotypic ratios for each cross.
- Discuss the significance of Mendel's use of mathematical and statistical analysis in his research.
- Why do you think Mendel was not immediately recognized as the discoverer of a new area of biology—genetics?
- Why does the gene interaction (incomplete dominance) NOT support the blending theory? How does it resemble the blending theory?

- What conclusions might Mendel have made if he had chosen *snapdragons* instead of *peas* for his study material?
- Why are the traits of human skin color and height not suitable for explaining the concept of simple dominance?
- There are four possible blood types in the ABO system. But how many *different* alleles are in the human population for this marker?
- What is the subtle difference between *incomplete dominance* and *codominance*?
- What is the significance of using upper- and lower-case versions of the same letter (e.g., *A* and *a*) for the dominant and recessive trait, respectively, rather than a capital *A* for dominant and the letter *B* (or *b*) for recessive?
- Assign the following paper to students and hold an in-class discussion/debate for and against genetic screening for diseases: Norrgard, K. (2008). Ethics of genetic testing: Medical insurance and genetic discrimination. *Nature Education*, *1*(1), 90.

Chapter 13 Possible Responses to Data Analysis Activities Questions

1. Approximately 600,000 bacteria enter normal cells and 80,000 enter mutated cells. Therefore, about 7.5 times more bacteria enter the normal cells.
2. Ty2 enters most easily. Approximately 600,000 Ty2 bacteria enter the normal cells compared to approximately 75,000 of strain 167 and 200,000 of strain 7251.
3. All of the strains have about an eight-fold reduction in number of bacteria that enter the cells. The differences in numbers of bacteria internalized into normal versus mutated cells are as follows: Ty2, 600,000 versus 80,000 (7.5-fold reduction); strain 167, 75,000 versus 10,000 (7.5-fold reduction); and strain 7251, 200,000 versus 25,000 (8-fold reduction). Thus, all the strains are similarly inhibited by the CFTR mutation.

Chapter 13 Homework Extension Questions

1. Black fur in mice (B) is dominant to brown fur (b). Short tails (T) are dominant to long tails (t). The genes for fur color and tail length are unlinked. What proportion of the progeny of the cross BbTt × BBtt will have black fur and long tails?
2. How many unique gametes could be produced through independent assortment by an individual with the genotype AaBBCCDdEe?
3. A cat has 66 pairs of chromosomes. Considering only independent assortment, how many genetically different kittens are possible from the mating of two cats? Is this number an overestimate or an underestimate?
4. Investigate the expression of two other traits in response to the environment.
5. How do UV rays accelerate melanin production?
6. What causes seasonal fur changes in the arctic fox?

CHAPTER 14: CHROMOSOMES AND HUMAN INHERITANCE

Chapter 14 Objectives

- Examine the rationale behind regional variations in human skin color.
- Analyze the role of a pedigree chart in analyzing human genetic diversity using examples.
- Analyze the inheritance patterns of autosomal dominant disorders and autosomal recessive disorders.
- Examine the different types of genetic disorders caused by X-linked inheritance patterns.
- Examine the different types of chromosome changes and their outcomes.

- Examine the ill effects of a change in human chromosome number using examples.
- Outline the applications of genetic screening and its potential benefits.

Enduring Understanding 3.C: The processing of genetic information is imperfect and is a source of genetic variation.

Essential Knowledge 3.C.1: Changes in genotype can result in changes in phenotype.

Chapter 14 Warm Up Questions

1. What are mutations?
2. How can a change in chromosome number occur?
3. How can changes in chromosome structure occur?

Chapter 14 Suggestions for Presenting the Material

- Students should be well grounded in their understanding of chromosomal structure before attempting to tackle the material in this chapter. The use of sketches, diagrams, and slides will greatly assist in making this material as clear as possible.
- They also must understand the events of meiosis, or they will have difficulty comprehending crossing over and changes in chromosome number resulting from nondisjunction.
- Remind students that crossing over and genetic recombination create variability among sexually reproducing organisms; challenge students to think about the role this plays in evolution.
- Describe the steps of Thomas Hunt Morgan's work to show how X-linkage was discovered. Ask students to solve the genetics problems that deal with X-linked genes at the end of the chapter. To assess how well students understand this material, work on as many of these problems together in class as time allows.
- Explain how karyotypes are prepared before showing a human karyotype. Otherwise, students might think that human chromosomes naturally occur paired up.
- Ask students to view some popular science videos (or show a short sequence) on possible future negative implications of genetic testing on newborns, such as in the movie *Gattaca*. This always brings forth lots of personal opinions and great questions from the students.
- Students need practice to learn how the different types of inheritance (autosomal recessive, autosomal dominant, and X-linked recessive) actually influence the inheritance of a trait in real-life examples. Review, if necessary, basic genetic terms such as *homozygous*, *heterozygous*, *dominant*, and *recessive*. To see how well students understand these types of inheritance, begin by working through some simple examples (as shown in Figures 14.2 and 14.7) of autosomal-dominant inheritance, autosomal-recessive inheritance, and X-linked recessive inheritance at the blackboard. Ask students to predict the possible phenotypic outcomes in each example.
- During lectures, use the genetics problems at the end of the chapter as they apply. Work through one or two examples at the blackboard with your class as a whole, and then ask students to complete the rest in class (possibly as part of a quiz) or on their own time. Students will enjoy the puzzle-solving aspects of pedigree analysis, while at the same time measuring their level of understanding of the different types of inheritance.
- Many of the genetic disorders and abnormalities mentioned in this chapter are ones whose names students have heard, but whose mechanisms of inheritance were unknown to them before reading this chapter. To lend more meaning to the conditions described here, ask students to think about the social and ethical problems associated with some of the diseases mentioned in this chapter.

Chapter 14 Classroom and Laboratory Enrichment

- Ask students (working individually or in small groups) to use chromosome-simulation kits (available from biological supply houses) to review chromosome structure, homologous pairing, crossing over, and independent assortment during gamete formation. If kits are unavailable, make your own chromosomes using a pop-it bead (or bead and cord) for each chromatid and a magnet for each centromere, or use the craft sticks mentioned in the Enrichment section for Chapter 10.
- Show karyotypes of males and females of different species without revealing the sex of the individual. Ask students to identify the sex.
- Discuss gene mapping in humans using the Human Genome Project's website to show some of the known locations of particular genes (http://www.ncbi.nlm.nih.gov/mapview/map_search.cgi). A free poster is available upon request to the Human Genome Landmarks Poster website (http://www.ornl.gov/sci/techresources/Human_Genome/posters/chromosome/).
- Project images of chromosomes with labeled gene sequences to demonstrate deletions, duplications, inversions, and translocations.
- Ask a local health unit or testing lab if you can copy (anonymously, of course) some karyotypes that show chromosomal defects. Show these to the class, and ask if the students can spot the defect before it is revealed to them.
- Prepare a slide to illustrate the relationship between DNA, genes, and chromosomes.
- Show the students unlabeled karyotypes from individuals with chromosomal abnormalities such as Turner syndrome, Klinefelter syndrome, or Down syndrome. Ask students if the karyotype appears normal; if not, what is wrong?
- Ask a genetic counselor to speak to your class about his/her job.
- Draw a pedigree for an unnamed genetic condition. Ask students if the disorder is autosomal dominant, autosomal recessive, sex-linked dominant, or sex-linked recessive.
- Obtain from your local health unit several brochures that explain the various genetic problems that can be inherited. Ask the students to discuss this information.
- From the same source mentioned earlier, you may be able to obtain a list of those genetic "diseases" for which there is mandatory testing in your state (usually of newborns). What voluntary testing programs are available?

Chapter 14 Classroom Discussion Ideas

- Why do harmful genes remain in the human population?
- If it becomes possible to easily and inexpensively choose the sex of your child, how will this change the male-to-female ratio among newborns? Do you think it is ethically correct to select the sex of your children? What about the effect of China's "one-child" policy on sex ratios?
- What is genoism? What will happen if we open Pandora's box and are able to read all the secrets found in DNA? What if we can tell how long your child could possibly live? What if you could genetically engineer your offspring (e.g., eye color, hair color, height, and elimination of all genetic disorders)?
- Do you think any of the traits Mendel followed in the garden pea were linked? Why or why not?
- If male and female offspring occur at a ratio of approximately 50:50, why do some couples have only boys or only girls?
- What is the distinction between the terms *gene* and *allele*?
- Why do individual chromosomes present at the conclusion of meiosis not have the same genetic constituency as they did before meiosis?

- If one sex of offspring tends to exhibit a trait more frequently than the other sex, of what is this an indication?
- What is the physical relationship of *genes* to *chromosomes* to *DNA?*
- What is the difference between a *translocation* of chromosomal segments and *crossing over*?
- Discuss the risks and benefits of amniocentesis. Would you elect to undergo this procedure (or urge your spouse to do so) if you had a history of genetic abnormalities in your family, or if you or your spouse were over 35? Why or why not?
- Why is hemophilia more threatening to the life of a female victim than to a male victim?
- Why do many people insist that girls cannot be red-green color blind (an X-linked inherited condition)?
- Can you think of some of the ethical questions involved in performing genetic research on humans? As our society learns more about genetic diseases, do you think couples who plan to have children should be required to have genetic counseling or undergo genetic screening?
- Why does the incidence of Down syndrome increase with the mother's age?
- When would you recommend genetic counseling to a couple, and when would you recommend genetic screening? Is there a difference?

Chapter 14 *How Would You Vote?* Classroom Discussion Ideas

- Assign students the following article on why race is not a biologically valid concept: http://www.sfgate.com/news/article/PAGE-ONE-No-Biological-Basis-For-Race-3310645.php#page-2.

  Monitor the voting for the online question. Like other species, genetic variation driven by natural selection is why we see variation in human populations across the globe. Skin color is one trait often used to categorize human populations into "races." From a biological perspective, race is biologically meaningless. Skin color variation is clinal, following patterns of the degree of exposure to UV and its effect on Vitamin D and folate levels. We know that individuals from the same geographic region will tend to be more genetically similar to one another, but there is more genetic variation within populations of humans from the same geographical areas than differences between populations. Given this information, ask students if they think that race is a meaningful biological concept.

Chapter 14 Possible Responses to Data Analysis Activities Questions

1. Kenya receives the most (354.21 UVMED). Ireland receives the least (52.92 UVMED).
2. People indigenous to Australia have the darkest skin color (19.30 skin reflectance). People indigenous to the Netherlands have the lightest skin color (67.37 skin reflectance).
3. In general, people indigenous to countries that receive higher amounts of UV radiation will have darker skin.

Chapter 14 Homework Extension Questions

1. If flower color shows incomplete dominance, what would the phenotype ratio of the F_1 generation be if you crossed a red flower with a white flower?
2. Can an O− female and an AB+ male have an A− child?
3. Investigate the Romanov's history with hemophilia? How is hemophilia inherited? Why was Alexis the only child to exhibit hemophilia?
4. How does nondisjunction occur at either anaphase I or anaphase II. What are possible results?
5. Give an example of a positive, negative, and neutral mutation. Explain the heterozygous advantage (also see Section 17.7) and give an example.

Chapters 11–14 Unit Closer

The development and action of the drug Taxol used to disrupt the growth of cancer highlights that application of student learning within this unit and provides an opportunity to apply and summarize the units learning. Students learned that errors in the regulatory mechanisms of cell division can lead to uncontrolled growth. The drug Taxol inhibits spindle formation in cancerous cells. This provides an opportunity for students to apply what they have learned about cellular activities, the necessity of division, and the need for control mechanisms.
Success Story: Taxol—dtp.nci.nih.gov/timeline/flash/success_stories/S2_Taxol.htm

Close with a short discussion as to what it means to be diagnosed with either a hyperthyroid or a hypothyroid condition; relate to cellular communication. Visit www.webmd.com.

Common Student Misconceptions

Students have trouble following the number of the chromosomes throughout the stages of meiosis. It is hard for them to understand when and how the chromosome number changes during meiosis? When is a cell haploid, when is it diploid? The students need to be well grounded in mitosis. Colored pipe cleaners can be used to represent paternal and maternal chromosomes. Have the students use the pipe cleaners to walk through the stages of meiosis I and meiosis II, showing duplicated sister chromatids, crossing over, alignment of homologues, separation of homologues, and two cells with still-duplicated sister chromatids. At meiosis II, they will be able to see that the pattern of the process is the same as mitosis.

Classroom Discussion and Activities

- Discuss how new discoveries in science can be incorporated into existing knowledge using Peter Agre's discovery of aquaporins and his subsequent Nobel Prize. Use the information at nobelprize.org to spark this discussion on the process of science and membrane structure.
- Discuss how our understanding of the structure of the membrane evolved into the current fluid mosaic model.
- Kidney dialysis lab available from various biological supply companies in kit form using unfiltered simulated blood.
- Bacteria identification and Gram staining lab—students can use dichotomous key to identify bacteria using their structure.
- The drug Taxol has shown to be effective in decreasing the growth of cancer cells. Have students research how this drug works at the cellular level and engage in a class discussion on its use in cancer patients and the possible impact on the trees used to harvest Taxol.
- The Cell Cycle and Mitosis Tutorial: https://www.youtube.com/watch?v=2aVnN4RePyI
- Meiosis tutorial: https://www.youtube.com/watch?v=16enC385R0w

Case Studies

The following case studies are from National Center for Case Study in Teaching Science (http://sciencecases.lib.buffalo.edu/cs/collection/):

- The Case of the Dividing Cell
- My Dog Is Broken
- You Are Not the Mother of Your Children
- The Case of Ruth James

AP® Practice Essays

1. State Mendel's two laws and
   (a) relate them to the process of meiosis.
   (b) apply them to crossing over, Down's syndrome, and sex-linked inheritance.
2. Describe how the phases of meiosis relate to Mendel's law of segregation and law of independent assortment.
   (a) Use a diagram to illustrate the movement and packaging of chromosomes into gametes.
   (b) Discuss how Mendel used his data from pea plants to predict offspring ratios.
   (c) What effect would nondisjunction of chromosomes have on the genetic makeup of gametes?
3. Genetic disorders may be caused by large-scale changes in the makeup of chromosomes.
   (a) Describe three events that can lead to a change in chromosomal structure.
   (b) Discuss how a karyotype could be used to identify such changes.
4. Explain the cell cycle of plants. Be sure to include a description of all the phases of the cell cycle.
5. Compare and contrast the processes of mitosis and meiosis in animals. Provide at least three similarities and three differences between the two.

Lesson Outline: Unit 3: Evolutionary Biology

Correlates with the 15th edition book, Chapters 16–19

- **AP® Biology Big Idea 1:** The process of evolution drives the diversity and unity of life.

Brief chapter summaries

Chapter 16 ("Evidence of Evolution") puts the development of evolutionary thought into historical perspective, highlighting much of the evidence that early evolutionary thinkers such as Darwin and Wallace relied upon to come to their conclusions.

Chapter 17 ("Processes of Evolution") gives a clear and concise description of the various ways that microevolutionary changes can happen and that may, eventually, result in various types of speciation.

Chapter 18 ("Organizing Information about Species") teaches students the various methods that people use to determine evolutionary relationships and how to visualize them, focusing on cladograms.

Chapter 19 ("Life's Origin and Early Evolution") explains the current understanding of how biological molecules, protocells, prokaryotic cells, and eukaryotic cells may have arisen on Earth, as best we know based on various types of experiments and observations that have been documented.

AP® Biology Learning Objectives 1.1–1.32 (Condensed and Summarized)

- Apply mathematical methods and conceptual understanding of natural selection and Hardy-Weinberg equilibrium to investigate the cause(s) and effect(s) of changes in the genetic makeup of populations over time.
- Explain and evaluate the evidence of evolution obtained from studies of fossils, morphology, biochemistry, physiology, and biogeography.
- Apply Darwin's theory of evolution to case studies and examples.
- Explain why genetic variations must be expressed phenotypically to lead to evolutionary change.
- Explain the Hardy-Weinberg Law and discuss its meaning relating to evolutionary biology.
- Explain what natural selection means in modern evolutionary theory and how natural selection may increase the frequency of a given allele in a gene pool of a population.
- Evaluate and create phylogenetic trees or cladograms that correctly represent evolutionary history and speciation.
- Describe how developmental biology and molecular biology provide insights into the evolutionary process.
- Explain the different hypotheses for the origin of life on earth. Support these ideas with scientific evidence.

CHAPTER 16: EVIDENCE OF EVOLUTION

Chapter 16 Objectives

- Examine the role played by asteroids in understanding the timescale of natural events.
- Examine how naturalists of the nineteenth century influenced theories about the natural world.

- Examine the new theories proposed by researchers about the evolution of the natural world.
- Explain the major observations of the natural world by researchers that influenced Darwin and his theory of natural selection.
- Examine the formation of fossils and their importance in studying the history of the natural world.
- Examine the role of radiometric dating in determining the age of fossil records.
- Discuss the characteristic features of the plate tectonics theory.
- Examine the major geological and biological events in Earth's history.

Enduring Understanding 1.A: Change in the genetic makeup of a population over time is evolution.

Essential Knowledge 1.A.1: Natural selection is a major mechanism of evolution.

Essential Knowledge 1:A.2: Natural selection acts on phenotypic variations in population.

Essential Knowledge 1.A.4: Biological evolution is supported by scientific evidence from many disciplines, including mathematics.

Enduring Understanding 1.C: Life continues to evolve within a changing environment.

Essential Knowledge 1.C.1: Speciation and extinction have occurred throughout the Earth's history.

Chapter 16 Warm Up Questions

1. What is the difference between evolution and natural selection?
2. What is the smallest unit that can evolve?
3. What is a species? How do you know?
4. How is variation in a population created?
5. Explain the classic example of pepper moth evolution.
6. Why is sexual reproduction the preferred method for most organisms on earth?
7. What are three pieces of evidence scientists use to support the theory of evolution?
8. How do you know when a population is evolving?
9. What is DNA and how is it related to evolution?
10. Why do some species go extinct and others don't?
11. Why did the dinosaurs go extinct 65 million years ago?

Chapter 16 Lesson Opener

Speedy: (10–15 minutes) Divide the class into two groups. Assign one group the task of supporting the teaching of evolution only in school and give the other group the task of supporting teaching *both* evolution and intelligent design. Give each group 10 minutes to prepare an argument. Have each group present to the class using a debate format.

Extensive: (40–50 minutes) Complete the speedy activity and follow up by watching the *NOVA: Intelligent Design on Trial*. http://www.pbs.org/wgbh/nova/evolution/intelligent-design-trial.html

As a follow-up assignment, have students pretend to be school administers and draft a letter that could be sent home to parents explaining why only evolution is taught in public high schools.

Chapter 16 Suggestions for Presenting the Material

- This chapter begins with a historical perspective. Present information with both a historical account and an emphasis on how emerging evidence has worked to change beliefs. Present as much evidence as you can. From this chapter, evidence from biogeography, comparative morphology, and geology (including fossil records, Earth history, and plate tectonics) are relevant.
- The National Science Education Center website (http://www.ncseweb.org/) has a variety of suggestions for presenting the evidence for evolution and links to other websites that have tutorials about evolution, fossil records, and more.
- Show students the movie "Flock of Dodos," which outlines the movement to supplant the teaching of evolution in classrooms with intelligent design.
- The textbook authors enumerate three areas (biogeography, comparative anatomy, and fossils) that were puzzling to early scientists. Use a learning cycle approach (http://www.narst.org/publications/research/cycle.cfm) to introduce each topic. Learning cycles involve three steps: 1) exploration, 2) concept invention, and 3) concept application. An example of how this might work with fossil evidence is as follows: 1) Students review pictures of potentially related fossils (or actual fossils) and get as much information about the fossils as possible. This could include radiometric or layering dates of fossils, locations of fossils, etc. 2) The students then try to draw conclusions about geological evidence as it relates to gradual change in the fossil record in order to define the concept of fossil evidence. 3) Finally, students work to apply that concept to other lines of, perhaps, less complete fossil lineages or multiple lineages in order to establish how fossil evidence supports evolutionary change.
- If you don't have time to investigate learning cycles, try to present the evidence first in order to allow your students to think about what the evidence means before they hear the synthesis about the evidence.
- If you need extra resources for geological information, there are two particularly informative websites:
  A. The American Geological Institute and the Paleontological Society published a short book entitled *Evolution and the Fossil Record* by John Pojeta and Dale Springer. An electronic version of the text is available online at http://www.agiweb.org/news/evolution/index.html. This manuscript has good descriptions of how geological evidence is used to interpret evolution. There is a discussion of several different isotopes used for radioisotope dating. There is discussion of fossil lineages with special emphasis on mammals, humans, and whales.
  B. The University of California has an online paleontology museum available at http://www.ucmp.berkeley.edu/. This is a very rich resource that has many other current external resource links. It also has sections that cover evidence for evolution from Earth's history.
- Before proceeding to the men who proposed a changing biological world, point out that the prevailing thought 200 years ago was "fixity of species." Because of the belief that species did not change, it was incumbent on humans to classify all living things. Although Linnaeus believed in this "fixity," his system is nevertheless still very valid and useful.
- One way to present the historical development of evolutionary thought is to chronicle the contributions of persons such as Lamarck, Lyell, Malthus, Wallace, and, of course, Darwin. Provide a spiraling narrative where you tell each individual's story concluding with their important evolutionary contribution. Finally, tie all of their work together by demonstrating how one individual influenced one or more of the others and how their cumulative work forms the basis for evolutionary theory. Set a historical timeframe; *On the Origin of Species* was originally published in 1859 at the time of the U.S. Civil War.

- Students rarely hear about Darwin's life other than his famous journey. Present his biography *before* his theory to spark interest. Perhaps the videotape listed in the Enrichment section could be used. You can then proceed to *natural selection* by first recalling *artificial selection* (maybe using dogs rather than pigeons as Darwin did in his book).
- Stress the tenuous conclusions that are in constant revision when scientists attempt to reconstruct the past. Just as a good medical diagnosis is not based on one examination or one lab test, a good analysis of past evolutionary history is not based on any one line of evidence, but rather several lines of corroborating evidence.
- You may wish to include a brief discussion of how radioisotopes are used to date fossils (Section 16.6). Certainly, emphasis should be placed on *how* these calculations are used rather than the actual calculations.
- Provide examples from comparative morphology of vestigial anatomical structures that are retained by organisms (e.g., pelvic bones in snakes, the human appendix, etc.).

Chapter 16 Classroom and Laboratory Enrichment

- Show a video describing Charles Darwin's voyage on the *HMS Beagle* and his thoughts as he traveled. The video series *Evolution* (2001, WGBH Educational Foundation) from PBS is excellent. Clips and other helpful material can be accessed at the following website: http://www.pbs.org/wgbh/evolution/
- Present fossil evidence (or 35-mm transparencies, filmstrips, or films of fossils) showing how a group of related organisms or a single genus (e.g., *Equus*) has evolved and changed through time. The students will most likely be surprised at the size of the horse's early ancestors. Pick a well-documented fossil lineage that shows many examples of gradual change or transitional states.
- If there is a museum of natural history near your campus, plan a field trip. The students would be able to view many relevant examples there to illustrate the concepts.
- Have a panel of other faculty from your campus come to class to serve as experts in the fields of paleontology, geology, Earth history, fossilization, comparative anatomy, animal diversity, zoology, botany, radiometric dating, anthropology, paleontology, etc. Have these "experts" prepare a brief explanation of their field's evidence for evolution, and then the panel can field questions from students.
- Whales, like snakes, have pelvic girdles. Show an overhead diagram of this portion of the whale skeleton.
- Generate interest in Darwin's theory by bringing a copy of *On the Origin of Species* to class. Read selected chapter titles and portions of the text. Point out the lack of illustrations in the original edition.
- Discuss methods of fossil preservation. Examine actual fossils or films, videos, or slides of fossils. Visit collections of fossils in nearby museums. Discuss why there are so many gaps in the fossil record due to the conditions required for fossilization.
- Prepare a summary table of the tools used in "reconstructing the past." For each tool, list the procedure, reliability, advantages/disadvantages, accuracy, and so forth.
- Show actual fossils or photographs of fossils representing some of the life forms prevalent during the Paleozoic, Mesozoic, and Cenozoic eras.

Chapter 16 Classroom Discussion Ideas

- This discussion may have started in Chapter 1, but have a discussion about how scientific knowledge is generated. What is the importance of evidence in generating a scientific theory? How is the use of the word theory different in the sciences than its use in other disciplines?
- What kinds of organisms are well represented in the fossil record? What types of organisms have left little or nothing in the fossil record? How does the process of fossil formation affect our knowledge of the history of organisms? Do the students feel there may be many species that are unknown, because they left no fossil record?
- How do worldwide distribution patterns of fossils suggest a common evolutionary origin for many organisms?
- What could fossils forming today tell scientists in the future?
- What was Jean-Baptiste Lamarck's contribution to our modern understanding of evolutionary theory?
- What steps did Lamarck fail to perform before setting forth his hypothesis of inherited characteristics? It seems obvious to us today that acquired characteristics are not inherited. As an example, if you have your nose altered by plastic surgery, do your offspring have the ability to inherit your original nose or the revised one?
- Does belief in the theory of evolution exclude belief in religion? Why or why not?
- This chapter focuses on many geological discoveries that are evidence for evolution. The American Geophysical Union and many other professional societies publish statements about the importance of evolution being a keystone theory in the sciences (http://www.agu.org/sci_soc/policy/positions/evolution.shtml). Why do professional societies make statements like this? Why is evolution so important when describing the current state of biological studies?
- Why was extensive travel a key ingredient in the development of Darwin's evolutionary thought? Do you think Darwin could have drawn the same conclusions if he had only studied the species in a single area?
- Can you think of any ideas commonly expressed today that are similar to Lamarck's understanding of evolution?
- How did the work of geologists such as Charles Lyell and others who were Darwin's contemporaries help Darwin create his principle of evolution?
- Those who wish to berate certain scientific principles sometimes say, "It's only a theory." This statement is used by those who oppose the theory of evolution. Does the use of "theory" in biology mean the concept is in doubt? Explain using examples.
- If you asked the following question in a sidewalk survey, what do you think the responses would be? "Darwin wrote a very famous book on the origin of ___________."
- Compare and contrast the principles of *uniformitarianism* and *catastrophism*. Evaluate the physical evidence for each.
- Research articles on "recent" human evolution. Are there any current trends in human evolution? Are there any structures that appear to be disappearing in humans?
- Is evolution a static idea that was proposed by Darwin and has never been modified or challenged since? What is the importance of the emergence of evidence of evolution from biogeography, comparative morphology, and geological discovery?
- There have been many sci-fi movies based on the premise that an entire organism, like a dinosaur, could be produced from a small preserved sample. What are the fallacies of such a proposition? Do you think this endeavor will be possible in the future?

Chapter 16 Possible Responses to *Critical Thinking* Questions

1. There is a chance that all carbon dating performed on the rock sample is accurate. Perhaps part of the sample was unearthed from a previous era approximately 225,000 years ago. The carbon dating procedure assumes that the material being tested has remained undisturbed since it was placed there. Since those specimens are located together now, they would yield different carbon dating results. Also, the carbon dating procedure assumes that all organisms utilize carbon 14 as equally as they do carbon 12. This is not always the case. There was an incident off the coast of Hawaii recently where scientists performed carbon dating on the shells of living animals. The results stated that the shells were over 2,000 years old! It is results like these that help us realize that there can be obstacles to overcome when attempting to determine the age of carbon-dated substances.
2. When comparing human history, even your own history, in a timescale that represents the entirety of Earth's history, the time we have been around is very short. If the most recent epoch (the Holocene) started about 12,000 years ago and that represents 0.1 second, then an individual's time on Earth might be close to 0.001–0.0001 second on a 24-hour clock.

Chapter 16 Possible Responses to *Data Analysis Activities* Questions

The iridium level in an average rock on Earth is only about 0.4 parts per billion (ppb). The average meteorite would contain around 550 ppb. This enables scientists to determine if the origin of a rock is from Earth or another source.

1. The iridium content of the K-T boundary layer is 41.6 ppb.
2. The iridium content of the K-T layer was 41.24 ppb more than the level of iridium at 0.7 m above the layer.

Chapter 16 Homework Extension Questions

1. In your own words, describe Darwin's theory of natural selection as the mechanism of evolution.
2. Explain how antibiotic resistance is an example of natural selection.
3. What ideas/people influenced Darwin in the development of his theory?
4. Explain an example of natural selection from nature.
5. How does artificial selection show human impact on variation in a population?
6. How does variation in a population occur?
7. In a theoretical population in Hardy-Weinberg equilibrium, 550 individuals have the *AA* genotype, 300 have *Aa*, and 150 have *aa*. For this population, $p = ?$ What are the five requirements for Hardy-Weinberg equilibrium?
8. Explain how fossils can be used to show the history of life on earth. Explain Hox genes and how they relate to evolution. Give an example of two mass extinctions and how they change the evolutionary history of life on earth?
9. Are we currently living through a mass extinction event? Why or why not?

CHAPTER 17: PROCESSES OF EVOLUTION

Chapter 17 Objectives

- Discuss the different factors contributing to microevolution.
- Examine the concept of genetic equilibrium under ideal conditions.
- Examine the three modes of natural selection—directional, stabilizing, and disruptive selection.
- Outline directional selection using three examples.
- Demonstrate the principles of stabilizing selection and disruptive selection using examples.
- Differentiate between sexual dimorphism and balanced dimorphism using examples.

- Examine the impact of genetic drift and gene flow on the genetic diversity of a population.
- Examine the process of reproductive isolation and the different methods of reproductive isolation.
- Examine the process of allopatric speciation using examples.
- Differentiate between sympatric speciation and parapatric speciation using examples.
- Distinguish the evolutionary processes of microevolution from macroevolution.
- Explain the different mechanisms of macroevolution.

Essential Knowledge 1.A.3: Evolutionary change is also driven by random processes.

Essential Knowledge 1.C.2: Speciation may occur when two populations become reproductively isolated from each other.

Essential Knowledge 1.C.3: Populations of organisms continue to evolve.

Chapter 17 Warm Up Questions

1. How are gene flow and genetic drift different?
2. What is a defining characteristic of the bottleneck effect?
3. How does a population's genetic diversity become reduced?
4. Explain the evolution of many different species of finches on the Galápagos Islands (not using scientific terms)?
5. What is a gamete and zygote?
6. Why are horses and donkeys still considered separate species when they can mate and produce mules?
7. Do all traits in a population change over time? Why or why not?
8. Does evolution have a goal? Why or why not?
9. Give an example of evolution happening today.

Chapter 17 Lesson Opener

Ask students to come up with the way in which natural selection worked in giraffe neck development from a shorter-necked similar organism. Many students will use language that implies a Lamarckian viewpoint that giraffes stretched their necks out of need. If this happens, ask them to come up with the molecular mechanism of how this stretched-out neck would get passed on to progeny. Get them to the point where they can see the Darwinian understanding that giraffes that happened to be born with alleles for longer necks were more effective in winning male–male fights for mates, and possibly have a selective advantage for accessing leaves high up in trees, despite the increased difficulty in getting water from the ground. The advantages made it such that giraffes with longer necks were more likely to have babies, and those babies would inherit the longer neck alleles.

Chapter 17 Suggestions for Presenting the Material

- The main points concerning variability and evolution are as follows:

  Variation is the result of several factors.

  The Hardy-Weinberg equation provides the baseline for calculating gene frequencies under unrealistic conditions.

  Several factors yield change in the real world:
  - Mutation
  - Genetic drift (founder, bottleneck)
  - Gene flow
  - Natural selection (stabilizing, directional, disruptive)

- Use artificial selection with domesticated dogs, for example, to introduce the concept of natural selection. Have the students research the incidence of certain conditions (e.g., hip dysplasia) in varying breeds of dogs. Can you see how these occur with more prevalence in particular breeds? Does this appear to be the result of inbreeding? Compare photos of various dog breeds to those of wolves. Can you see the huge variation in the appearance of different breeds even though they are derived from a common ancestor?
- Work with the students to develop a visual model that demonstrates how alleles will be selected for over time. Draw a picture with two rectangles representing a population's gene pool at two different times. Develop a model that uses the various microevolutionary processes (survival, sexual selection, gene drift, and gene flow) to select for alleles in a population over time. Develop several scenarios that could be used to demonstrate the model. For example, one reason an allele is selected for is that it allows for better survival. Point out that giraffes with longer necks were selected for because the longer neck ensured they would have a better food supply, despite the disadvantage that having a long neck is for drinking water. Peacocks provide a good example of sexual dimorphism caused by sexual selection.
- Natural selection is, in many scientists' views, the most important concept for biology students to understand. Therefore, it is vital that you clear up any student misconceptions by providing many clear examples of microevolutionary processes. Don't just explain examples; have students work through examples. Provide the students with examples of where selection might/should occur and have the students describe how selection would work in each scenario. As the students describe their scenarios for the class, use their misconceptions to point out more correct interpretations.
- Emphasize to students that evolution occurs in populations, not individuals, even if they can see the changes in an individual's phenotype. Stress that selection pressures result in the survival of some individuals in the population over others. Use real-life current examples of evolution in action (e.g., HIV, Ebola, and MRSA).
- Work several problems with the Hardy-Weinberg formula to demonstrate allele frequency.
- The different results of natural, stabilizing, directional, and disruptive selection are fairly easy to picture when looking at the butterfly graphs in the chapter. Try to help students conceptualize how each type of selection works using multiple examples from the text and from other sources. If you develop a model for microevolution as described earlier, have students apply scenarios with the model and provide enough information for natural selection examples that students should also be able to predict how selection will work in the sample scenarios. Several basic scenarios are presented as a starter next.

  Sample Scenario A—Stabilizing selection: Sharks are natural predators of seals. Seals have a variety of body-size phenotypes. Sharks tend to attack smaller seals because they struggle less. Larger seals have a harder time escaping.

  Sample Scenario B—Directional selection: A new bird species migrates to a volcanic archipelago. Males of the species have several phenotypes for color ranging from a light brown, to a medium brown, to a dark brown. On the island, females show a preference for darker-brown males.

  Sample Scenario C—Disruptive selection: A finch species shows continuous variation in beak width. Birds with smaller beaks feed on smaller seeds, birds with medium beaks eat medium-size seeds, and birds with large beaks eat large seeds. A drought occurs that affects the plants that produce the medium-size seeds the most.
- Pesticide resistance and antibiotic resistance are both fairly accessible and well-documented examples of natural selection that have occurred on observable timescales. Recent MRSA (methicillin-resistant *Staphylococcus aureus*) outbreaks have propelled bacterial resistance into newspapers and magazines. Present a range of different scientific and popular articles that document MRSA. Spend time discussing why antibiotics and antibacterial soaps create the conditions that lead to selection in bacteria.
- Discuss the patterns for macroevolution. Show documented examples of coevolution and examples of adaptive radiation following migration to a new location or after mass extinctions.

Chapter 17 Classroom and Laboratory Enrichment

- **AP® Investigation #1: Artificial Selection**
- **AP® Investigation #2: Mathematical Modeling: Hardy-Weinberg**
- Choose an easy-to-see trait governed by one gene with two alleles, such as earlobe attachment (earlobes whose bases are not attached to the jawline) or tongue rolling (the ability to roll your tongue without holding it with the sides of your mouth), and ask students to determine their own phenotype. Determine the number of homozygous recessive individuals (*aa*) in the class. Use the Hardy-Weinberg principle to calculate the frequencies of the dominant allele and the recessive allele. Examples: earlobe attachment (unattached–dominant, attached–recessive), hairline (widow's peaks–dominant, straight–recessive), freckles (freckles–dominant, no freckles–recessive), and (tongue rolling–dominant, no tongue rolling–recessive).
- Demonstrate genetic drift by tracing changes in allele frequency through time. In small hypothetical populations, select a trait governed by two alleles and calculate the frequency of each allele. Different groups of students could be assigned populations of different sizes. Follow each population through several generations, as some of its members (selected by coin tosses) succumb to disease, predation, and other random causes of early death. How does population size affect genetic variation over time?
- Show slides or videos about endangered species that are threatened by sharp reductions in population size and subsequent loss of genetic variability due to genetic drift.
- What happens to the genetic variability of small, isolated populations of laboratory organisms after many generations without the introduction of new organisms? Design and implement an experiment using any organism with a short generation time and several easy-to-see traits that can be followed from one generation to the next.
- Find a good natural selection tutorial online (e.g., http://www.biologycorner.com/worksheets/pepperedmoth.html has a good peppered moth example), and have students collect and analyze data using the simulation. This simulation could be used to generate numbers for the Hardy-Weinberg formula.
- How does artificial selection by humans affect gene frequencies of domestic plants and animals? Pursue this question with experiments or demonstrations.
- Select a well-known example of the founder effect in a human population, such as the Amish population with Ellis-van Creveld syndrome. Using slides, show how a phenotypic characteristic (e.g., polydactyly) "spreads" from one generation to the next. Are there any other human populations that illustrate the founder effect?
- Explain the development of insect resistance to DDT as a modern-day example of natural selection. Point out that DDT was introduced to the world in the early 1940s, and in just 10 years, resistant strains were reported in many countries. By the time it was banned in the United States in 1973, virtually every housefly was resistant to its effects. How did DDT use impact other organisms?

Chapter 17 Classroom Discussion Ideas

- How could the results of various scientific experiments be influenced by using strains of laboratory mice that are continuously inbred?
- How might antibiotics and antibacterial soap affect bacterial populations, especially pathogenic populations?
- Since antibiotics only kill bacteria, they should not be used to treat viral infections. How might the widespread use of antibiotics for any type of infection lead to the development of antibiotic-resistant strains of bacteria?
- What is artificial selection? How does it differ from natural selection?

- How does sexual selection benefit a species? Would the introduction of alleles from a similar but different species introduce variety and thus help the species? Why or why not?
- What is speciation? How does speciation occur? When might you expect to see speciation?
- What is the difference between microevolution and macroevolution? What are the different types of evidence that support each of those concepts?
- How does phenotypic variation arise? Ask students to list as many sources of phenotypic variation as they can. They should be able to remember how genetic variation comes about from their earlier study of genetics.
- What effect would phenotypes altered by environmental conditions have on future generations (e.g., plant color affected by soil pH)?
- How are new alleles created? Is the creation of new alleles an important source of genetic change? Why or why not?
- Why doesn't evolution lead to perfect organisms? Think of the trade-off with giraffe necks, the benefit of being selected as a mate, food foraging at the risk of drinking, or sexual selection that leads to bright bird plumage that is detrimental to blending in with the environment.
- Although not a pleasant topic, many scientists predict that another mass extinction will take place. Review the arguments presented by these researchers.
- What are some phenotypic variations that might have assisted the success of *Homo sapiens*? Ask your students to think of some imaginative examples of variations that might be useful in the future evolution of our species.
- Think of examples of human alleles whose frequencies vary from one global region to the next.
- What did Darwin's study of the different finch species among the Galápagos Islands tell him about speciation? What conclusions can you make about the evolutionary histories of the different species of Galápagos finches, given what you now know about the process of speciation?
- How does sexual selection benefit a species? Would the introduction of alleles from a similar but different species introduce variety and thus help the species? Why or why not?
- Which of the factors that cause changes in allelic frequencies could be under conscious human control?
- Some conservation groups are working to preserve the genetic material of endangered species (http://www.msnbc.msn.com/id/5526627/). Why might bringing these animals back from extinction using limited samples from the population be detrimental to rescuing those populations?
- Why is the statement "She has evolved into a fine pianist" *not* biologically accurate? Would a great tennis player, like Venus or Serena Williams, have children who are "naturally gifted" tennis players?
- Of the five sources of phenotypic variation, why is *mutation* the only one that *creates* new alleles? How do the others yield variation? What is the importance of the variation?
- How could the results of scientific experiments be influenced by using strains of laboratory mice that are continuously inbred?
- How might bacterial-resistant genes in nonpathogenic strains of bacteria affect us in the future? What might be the effect of releasing herbicide-resistant genes or genes that alter the ripening schedule of fruits into commercial plant crops?
- Research the coelacanth. Why was its rediscovery so important to scientists?
- Have your students research articles on drosophila from the Hawaiian Islands. Do they serve as an example of sympatric speciation?

Chapter 17 Possible Responses to *Critical Thinking* Questions

1. One problem with the terms *primitive* and *advanced* is that they reflect value judgments and are not scientifically or biologically meaningful. The biggest problem with such terms is that they imply evolution is directed toward progress or advancement toward a particular goal or purpose. A more accurate comparison between two species might be to say that a trait is ancestral in one species and derived in another, representing their relationship relative to each other.
2. Rama the camel has survived the prezygotic mechanisms for reproductive isolation insofar as a cross between a camel and a llama has produced a successful hybrid organism (for pictures, check out http://www.taylorllamas.com/Camel-LamaCrossPhotos.html). Most hybrid organisms, however, are sterile. It is thought that llama and camel hybrids may be able to produce viable offspring because of the compatibility of their chromosome number (74).
   If Rama were able to successfully reproduce, it would be evidence that camels and llamas are the same species. Mayr's definition of a species included the need for interbreeding. This may be an artificial designation, as geographically separated species may still be able to reproduce; the physical barrier just prevents that from happening.
3. According to the biological species concept, a species is defined as where the parents not only produce offspring, but that those offspring must be capable of successful reproduction themselves. Therefore, the two species of antelopes cannot be classified as a single species until they produce viable progeny.
4. If it is true that sexual selection is responsible for some uniquely human traits, then a simple experiment should be able to demonstrate a pattern of selection that favors these characteristics.
   If it is true that female selection has led to the selection of "charming, witty" personalities in males, this preference should hold up in sample populations of females. For example, several men could be selected to be videotaped displaying social behaviors defined as "witty" and "charming." In the first experiment, the same male would portray a range of "male" behaviors including some deemed charming and witty. When asked which behaviors the females preferred, they should select the behaviors that are identified as charming and witty a majority of the time. In a similar experiment, two males would be videotaped, one who is more attractive, but less witty and charming, and another who is more charming and witty but less attractive. When video of these two archetypes are exhibited for females, females should choose witty and charming as a potential mate more frequently than physical features.
   A similar experiment could be conducted in male populations. In males, there should be a selection preference for less hair and softer features. Male subjects would be asked to rate photos of females that have been altered to show the same individuals with varying levels of hair and varying levels of "softness" in their features. The results should demonstrate a preference for the previously identified features.

Chapter 17 Possible Responses to *Data Analysis Activities* Questions

There is a constant battle between rodents and pesticides. Often rats that develop immunity to pesticides increase in the population.

1. The town with the most rats susceptible to warfarin was Ludwigshafen.
2. Stadtlohn had the highest percentage of poison-resistant wild rats.
3. In Olfen, 79% of the rats were warfarin-resistant.
4. It appears as though Stadtlohn had the most extensive applications of bromadiolone.

Chapter 17 Homework Extension Questions

1. How do populations become small enough for natural selection to work on?
2. Give an example of how genetic variation decreases in a population.
3. Why does gene flow typically occur more in mobile organisms?
4. Pictorially represent allopatric and sympatric speciation.
5. Name four prezygotic barriers and how each works?
6. The swallowtail butterfly provides a potential example of sympatric speciation. In this case, female butterflies lay their eggs on parsley plants, and the caterpillars develop and metamorphose into adults, at which point they mate with each other and the females fly off and deposit eggs on a new parsley plant. These populations originated from a species that feed on carrot plants. A carrot plant might be right next to a parsley plant, but because of the egg-laying preference difference, butterflies of each population never meet to mate with each other. The change in preference appears to have been caused by a genetic mutation that is passed from one generation to the next. Explain what sympatric speciation is and how this might be an example of sympatric speciation. Describe the process by which these two new populations will become two new species.
7. How does evolution explain both unity and diversity?
8. Give an example of how humans continue to evolve?
9. What happens to a species that can no longer evolve? Why might this occur?

CHAPTER 18: ORGANIZING INFORMATION ABOUT SPECIES

Chapter 18 Objectives

- Discuss the impact of non-native plant and animal species on the honeycreeper species.
- Examine the importance of phylogeny in understanding the shared ancestry of all species.
- Differentiate between the mechanisms of morphological convergence and morphological divergence.
- Outline the role played by biochemical molecules in influencing the evolution of different species.
- Examine how homeotic genes influence and ensure optimal developmental patterns in organisms.
- Examine the applications of phylogeny in conservation biology and medical healthcare.

Essential Knowledge 1.B.2: Phylogenetic trees and cladograms are graphical representations (models) of evolutionary history that can be tested.

Chapter 18 Warm Up Questions

1. How can you tell how closely related two species are?
2. Name a trait that is present in two groups of animals but did not evolve from a common ancestor.
3. What is a possible out-group for a cladogram of fish?

Chapter 18 Lesson Opener

The website, http://biology.fullerton.edu/biol261/phylolab.html, presents an excellent tutorial on the interpretation of cladograms. Divide the class into small groups and have each discuss one of the cladogram questions on the website. Performing this type of practice exercise will help in the overall interpretation of the chapter.

Chapter 18 Suggestions for Presenting the Material

- Provide historical perspectives on taxonomy. Start with Linnaeus and show how the same organisms can be grouped differently based on the characters used for analysis.
- Spend time making the distinction between divergence/homology and convergence/analogous structures.
- Provide other examples of convergence and divergence, homology and analogy. The book has several good examples, but there are others like mammal teeth patterns (divergence), or spines or thorns on plants (convergence).
- Explore the comparative analysis of cytochrome *b*. Simplify that example and have your students compare the similarities.
- Modern photography of embryos at comparable stages does demonstrate developmental similarities. Use this photographic evidence to have a discussion about how small changes in development plans could lead to large morphological changes over time.
- Biochemical evidence of evolution is a current and expanding body of work. Use contemporary examples of how protein and DNA analysis is elucidating evolutionary lineages.
- Start with fairly straightforward examples of organisms that are easily classified and then work up to more difficult examples like dolphin/whale or platypus lineages.
- Use the example of the foot of varying mollusks to provide another example of homologous structures.

Chapter 18 Classroom and Laboratory Enrichment

- **AP® Investigation #3: Comparing DNA Sequences to Understand Evolutionary Relationships with BLAST**
- The Tree of Life Project (http://tolweb.org/tree/phylogeny.html) has very detailed phylogeny trees and some good instructional tools for completing trees.
- Provide your students with other examples of DNA or protein sequences and have them analyze the data and determine how similar organisms are.
- *Hox* is a well-documented gene in *drosophila* development and *drosophila* relationships. *HOM/Hox* expression has been studied in regard to segment formation in arthropods. Use comparisons in these lineages to demonstrate how master gene mutation can lead to dramatic morphologic change.
- This chapter comes after the chapters on genetics but before the chapters on development. Perhaps having a brief primer lesson on development patterns will help set a frame of reference for developmental changes.
- Have students construct evolutionary trees using one character at a time, then two, then more. Have students reflect on the changes that occur in their trees when including more characters. Have students compare their trees to those done by taxonomists.
- Compare how an evolutionary tree is analogous to a family tree or genealogy to help students better understand phylogenetic trees.
- Recent genomic analysis of the platypus is shedding light on its common ancestry. Present and discuss evidence used to determine where the platypus fits into the animal kingdom (http://www.nature.com/news/2008/080507/full/453138a.html, with video interview and podcast).
- Honeycreepers are also indigenous to Brazil. Are some of the species there also endangered? How are the Brazilian birds different from those in Hawaii?
- Research the literature for other examples of phylogeny based on character analysis. This website, http://www.pnas.org/content/97/21/11359.figures-only, shows some figures from a study comparing the mitochondrial DNA of flatworms. Look for other similar examples on the Internet.

Chapter 18 Classroom Discussion Ideas

- What are some of the different methods you could use to classify the diversity of the honeycreeper species? What would be the best way to demonstrate evidence that the honeycreepers are descended from a single common ancestor?
- The platypus presents a variety of traits that we would consider novel for a variety of species that might not seem closely related. How could we work to classify the platypus? What will provide stronger clues, morphological status, or genomic analysis?
- What are the different methods that phylogeneticists use to classify evolutionary change? How have these methods changed over time?
- For each of the following pairs of structures, have your class determine if they are analogous or homologous: moth wing and bird wing, shark fin and dolphin fin, and octopus eye and human eye.
- Show graphics of fetal and adult skulls of different mammals. Do they show the same proportional changes as the skulls of humans and chimpanzees?
- How has taxonomy changed since Linnaeus's time?
- Why is evolutionary cladistics today so difficult in an evolutionary timescale context? What type of evidence would be required to revise a cladogram?
- How does the human factor come into play when reconstructing evolutionary history?
- Discuss the benefits and problems associated with regional naming of organisms?
- How will biochemical analysis differ if the focus is on DNA or proteins? Which (DNA or protein) provides a more consistent basis for analysis? Which will be more conserved in related lineages?
- How does analysis of 20 species versions of cytochrome *b* provide evidence for common lineage? What does protein analysis tell us about common ancestry?
- What are some of the different methods of comparing proteins or comparing DNA, and how can the methods be used to demonstrate evolutionary change?
- Comparative genetics has demonstrated that humans are roughly 1–2% different from chimpanzees, and that there is about 60% genetic conservation between human and fruit flies. Even more humbling is the comparison between humans and a banana! Why are these seemingly different organisms so genetically similar?
- What does the evidence from vertebrate forelimbs tell us about evolutionary history?
- How do the lineages of vertebrate forelimbs and the lineages of animals with wings tell us different evolutionary histories? How are the two patterns of development different?
- Why would species that look so different at an adult stage look so similar during embryonic development?
- Why would a master gene or a homeotic gene mutation be either so lethal or change the body plan outcome so drastically?
- What does the evidence obtained during *Hox/Dlx* experiments tell us about homeotic gene expression that may have played an important evolutionary role?
- What are the differences between lethal, neutral, and beneficial mutations? What would actually lead to a mutation being lethal, neutral, or beneficial?
- What is a molecular clock and how is it calibrated?
- Give a detailed presentation of the axolotls salamander. What juvenile features can the students identify?
- A new primate species is discovered. If you were a taxonomist, describe the steps you would take to classify the new species and determine its closest relative.

- Describe how bacteria are classified and how it is different from how other organisms might be classified.
- Detail the adaptation of the horse hoof as an example of morphological divergence.
- Describe the difference between morphological divergence and convergence, including a discussion of homologous and analogous structures.
- Why does DNA and protein analysis provide a much more distinct evolutionary picture than morphology?
- Mitochondrial DNA provides some fairly clear evidence of human evolutionary lineage. Describe the benefit of studying mitochondrial DNA and detail the evidence provided by the genomic study of mitochondrial DNA.
- The Human Genome Project was an ambitious program to study the entire human genome. Several other genome projects have been completed and others are on the way. What can we learn about evolutionary lineage from these types of studies?
- What types of genes are highly conserved between species? Why does the conservation of those genes end up being important to survival?
- Compare and contrast the development of several closely related species. Where are the changes in their developmental plans? What leads to those changes?
- Recent work has elucidated a genomic comparison of the platypus. The platypus is an incredibly unique mammal that retains characteristics of birds and reptiles. It has venom and a reptilian gait, lays eggs, and has hair. The platypus bill is a sensory organ. While the platypus has mammary glands, it lacks teats like other mammals. How has genomic analysis led us to a better understanding of where the platypus fits into an evolutionary lineage?
- Research the early development of the idea of a "molecular clock."
- Compare the biochemical analysis of several seemingly unrelated species. Based on the data, which appear to be the most closely related?
- Compare photographs of the embryos of many different species. Which have similar characteristics? Has the advent of the study of embryology influenced how scientists have reclassified different species?
- Look into the studies of viruses that appear to "jump" from infecting one species to another.
- Research the axolotls salamanders. Discuss how their juvenile features appear to be advantageous to the adults of the species.

Chapter 18 Possible Responses to *Critical Thinking* Questions

1. The phrase "ontogeny recapitulates phylogeny" sounds quite pretentious and may not immediately make sense. In a nutshell, the term refers to the fact that there are structures present on a human embryo that would have been present in our ancestors. For example, a human embryo has a yolk sac, gill slits, and a tail. Obviously, these features are not present in an adult human! They are evidence of traits that our ancestors possessed and, thus, show an insight into our evolution as a species.
2. Synpolydactyly is a disorder that leads to the development of webbed fingers. In order for this disorder to arise, a mutation would have to occur in the set of master genes that controls finger development. Scientists have discovered the precise Hox gene that, when mutated, results in synpolydactyly. Mutations at this site can also cause other limb abnormalities, such as brachydactyly (shortened fingers and toes).

Chapter 18 Possible Responses to *Data Analysis Activities* Questions

Related species may be identified using bone morphology and the analysis of DNA sequences. The honeycreeper clade was analyzed using these four criteria.

1. The species that represents the out-group is the common Rosefinch.
2. According to the cladogram, it appears that the Akohekohe is the most closely related to the Apapane.
3. According to the cladogram, the sister group to the Akikiki is the Maui Alauahio.
4. The Liwi is more closely related to the Palila than the Maui Alauahio.

Chapter 18 Homework Extension Questions

1. Draw a cladogram that explains the relationship between sharks, birds, amphibians, fish, reptiles, and mammals. Be sure to include when the characteristics evolved (indicate using a hash mark across the line).
2. Why do evolutionary biologists use cladograms to show evolutionary history?
3. Explain the rule of parsimony by creating four different cladograms with species A, B, C, and D represented. Which tree is most parsimonious (most likely to occur in nature) and why?

CHAPTER 19: LLIE'S ORIGIN AND EARLY EVOLUTION

Chapter 19 Objectives:

- Examine the significance of ozone layer in sustaining life.
- Examine the formation of the Earth from the explosion that formed the universe according to the big bang theory.
- Examine the different theories that explain the formation of organic molecules.
- Examine the evolution of a protocell from organic monomers using a flowchart.
- Examine the evolution of bacteria and eukaryotes using examples.
- Examine the role played by membranes in distinguishing eukaryotes from bacteria and in the formation of the endosymbiont hypothesis.

Enduring Understanding 1.B: Organisms are linked by lines of descent from common ancestry.

Essential Knowledge 1.B.1: Organisms share many conserved core processes and features that evolved and are widely distributed among organisms today.

Enduring Understanding 1.D: The origin of living systems is explained by natural processes.

Essential Knowledge 1.D.1: There are several hypotheses about the natural origin of life on Earth, each with supporting scientific evidence.

Essential Knowledge 1.D.2: Scientific evidence from many different disciplines supports models of the origin of life.

Chapter 19 Lesson Opener

Which came first, the chicken or the egg? That may not be a great evolutionary question; chickens were part of a group of animals that lay eggs. But here's one that you should actually take some time to ponder: Which came first: proteins, RNA, or DNA? What evidence is out there that supports any of those lines?

Chapter 19 Warm Up Questions

1. Name a feature humans share with a common ancestor. Why is this true?
2. How do we know all organisms evolved from a common ancestor?
3. Bird and bat wings are an example of what idea?
4. What molecules were present on early earth?
5. What molecule was noticeably absent from early earth?
6. What is the difference between monomers and polymers?
7. What is the big bang theory and what does it explain?
8. What is the difference between a prokaryotic and eukaryotic cell?
9. What is the function of a plasma membrane and why do cells need it?

Chapter 19 Suggestions for Presenting the Material

- This chapter continues the theme of evolution (Unit III) but "steps back" in time to look at the origin and early evolution of life in preparation for the survey of life forms in Unit IV.
- Stress the tenuous conclusions that are in constant revision when scientists attempt to reconstruct the past. Just as a good medical diagnosis is not based on one examination or one lab test, a good analysis of past evolutionary history is not based on any one line of evidence, but rather several lines of corroborating evidence.
- Present the evidence. For example, the chapter talks about the Miller experiments that led to the formation of sugars and amino acids. Opponents of evolution often cite that scientists do not agree that Miller's conditions were accurately based on more contemporary analysis. However, many other studies have been done that replicate similar results under different conditions (http://ncse.com/creationism/analysis/icon-1-miller-urey-experiment for a summary). Present this evidence. Miller continued to work on this problem. There is a good Web-based simulation of his technique (http://www.ucsd.tv/miller-urey/) with a video interview about his ideas and video that demonstrates the real experimental apparatus.
- Many current hypotheses about molecular origins focus on RNA as the important first molecule that could have led to DNA and proteins. Present evidence for many of the RNA-based hypotheses.
- Provide the students with data sets or simulations and help them make claims based on the empirical evidence.
- Because of the rather theoretical and speculative nature of the "origin of life" section, students' minds often wander because they don't have any concrete terms to write down. Use of visual material of any kind will help to retain focus.
- Note that The Timeline for Life's Origin and Evolution in Section 19.6 presents an evolutionary tree of life using a timescale devoid of the usual era names.

Chapter 19 Classroom and Laboratory Enrichment

- Prepare an overhead or slide summarizing the work of Redi, Spallanzani, and Pasteur in disproving spontaneous generation under "recent" conditions on Earth.
- Have students do reports on some of the different building block experiments since Urey/Miller. Have them summarize the outcomes and the likelihood of the scenario they study.
- Present evidence of current organisms that live in conditions on Earth that are extreme and how they have adapted to those conditions. Pick organisms that live in hot springs, thermal vents, or deep in the soil and Earth's crust.
- Describe some of the key cellular adaptations in a timescale relationship.

- Lay out the evidence that helps describe the endosymbiotic theory presented by Dr. Lynn Margulis (see Berkely's page summarizing her theory: http://evolution.berkeley.edu/evolibrary/article/history_24).
- Show actual fossils or photographs of fossils representing some of the life forms prevalent during the Paleozoic, Mesozoic, and Cenozoic eras.
- Discuss the locations of hydrothermal vents. There are distinct examples of food webs that exist in these unique habitats.
- Talk about the chemical composition of recovered meteorites. Has there been any indication that these meteorites contain the precursors for life?

Chapter 19 Classroom Discussion Ideas

- Why do you think the Archaeans are able to live in so many different extreme environments?
- Evolutionary biologists believe that life on Earth first evolved in the absence of oxygen. Where did the oxygen in our atmosphere come from?
- Would you consider nanobes to be living organisms? How are they like protocells?
- Can you think of any environment on Earth within which life cannot exist?
- Astrobiologists doubt the presence of life above ground on planets without an ozone layer. What are the ramifications for life on Earth if our ozone becomes damaged further?
- Make a poster-sized presentation of the Earth's timeline. Include photos obtained from the Internet to accompany your presentation.
- On what evidence do scientists reconstruct the primordial Earth depicted in Figure 19.4?
- Why was no free oxygen (O_2) included in Miller's experiment? What did the electrical spark simulate?
- What distinguishes a nonliving lipid-bound sphere containing nucleic acids and amino acids from a living cell?
- Name some organisms living today that are virtually unchanged from their earliest appearances in the fossil record millions of years ago. Why do some organisms fail to change significantly throughout geologic time?
- In a hypothetical series of events leading from spontaneous formation of molecules to living cells, what do we know about how each step may have occurred?
- Why was aerobic respiration necessary for the evolution of eukaryotes?
- Describe the metabolic pathways used by the first living cells to obtain energy.
- What are some of the advantages of multicellularity? What are the advantages of a nucleus?
- Why do you think people believed as recently as 100 years ago that spontaneous generation still occurs on the Earth?
- The branching of prokaryotes during the Archaean into archaebacteria, eubacteria, and eukaryotes is the basis for a new three-domain scheme of classification. Evaluate the merits of this scheme versus the more traditional five- (or six-) kingdom scheme.
- Look at the timeline scale in Section 19.6. Prokaryotes were around for a long time. What are some of the key adaptations that led to the radiations and origins of plants, animals, and fungi all around the same time?
- Why are extinctions important in the development of new species?

- Remember back to the last chapter where you read that some common ancestor diverged into bacteria and archea. Then the archea lineage diverged into eukaryotes. This chapter details how cyanobacteria, from the bacteria lineage, developed the early photosynthetic metabolic pathways. How is it possible that all plants, which are eukaryotes, are somehow related to bacteria that emerged about one billion years *after* the bacterial and eukaryotic lineages diverged?
- What evidence exists that makes scientists conclude that the red algae was probably the first eukaryote?
- Discuss the history of the experiments by Stanley Miller.
- See if you can locate the original article by Miller and Urey in which they reported their famous experiment. Read it carefully to see what speculative application they made for their experiment.
- Although the idea of life originating on Earth from forms traveling from distant planets is highly speculative, there is evidence that exists that may support this hypothesis. Find the evidence and describe why it may be credible.
- Discuss experiments that have been conducted since Miller in the 1950s and describe how they help to contextualize Miller's findings.
- Discuss recent controversial studies that use human mitochondrial DNA to draw conclusions about human evolution. Based on what you find about mitochondrial origins and reproduction and the endosymbiotic theory, describe the evidence that the mitochondria provides for early evolution.
- Research the laboratories that are currently investigating life on early Earth. What advances have they made from what has been presented in Chapter 19?
- The United States is working to incorporate future trips to Mars into its space program. Recent discovery of what may have been water on Mars is fueling the impetus for these trips. What would the significance of finding water or evidence of past water on Mars mean? What would it mean if we found evidence of life on other planets?
- Research articles about life in the Atacama Desert (depicted in Figure 19.1). What are the drastic conditions there that seem almost incompatible with life?
- Discuss the research by Jack Szostak that relates to the study of protocells. What conclusions were reached as a result of his experimentation?

Chapter 19 Possible Responses to *Critical Thinking* Questions

1. There are two main reasons that fossils older than three billion years would not exist. First, there would not be much to fossilize and so the conditions would have to be absolutely perfect. Second, because of plate tectonics, the layers containing those organisms may not exist anymore. Not only would those fossils be at the deepest layers of sedimentation, but those layers may have been at the edge of where two plates meet. Those layers may have been destroyed by friction or returned to a deeper level and turned into molten magma.
2. Carl Sagan said this because all of the ingredients for life as we know it (e.g., carbon, magnesium, and calcium) are found in the interior of stars. Stellar explosions then release these chemical elements.
3. The oxygen-releasing photosynthetic activity of early cyanobacteria is thought to have caused the mass extinction of anaerobic organisms. It is hypothesized that some anaerobic bacteria ingested aerobic bacteria resulting in a symbiotic relationship.

Chapter 19 Possible Responses to *Data Analysis Activities* Questions

Over many billions of years, asteroids have impacted the composition of the atmosphere.

1. A decline in asteroid impacts occurred before the atmospheric oxygen increased.
2. Since the first cells arose, the level of carbon dioxide has dropped and the level of oxygen has risen.
3. In today's world, oxygen is more abundant than carbon dioxide.
4. The rise in carbon dioxide at the far right of the graph is likely due to the burning of fossil fuels by cars and industry.

Chapter 19 Homework Extension Questions

1. If species 1 and 2 have similar appearances, but different DNA sequences, while species 3 and 4 have different appearances, but very similar DNA, which pair of species is more likely to be closely related? How can this be explained?
2. Based on your knowledge, describe four features of a common ancestor for all living organisms.
3. How can it be explained that snakes and lizards are both reptiles but have different features. Explain the steps outlined by Oparin and Haldane for the origin of life.
4. Explain the Miller and Urey experiment? Why is this experiment no longer accepted as what happened to create organic molecules?
5. Why were the first cells on early anaerobic prokaryotes? How did aerobic cells evolve? Pictorially represent the endosymbiont theory and give three pieces of evidence that support the theory?
6. What are the similarities and differences between archaea and bacteria?
7. What are cyanobacteria and why were they important to the formation of life as we know it today?

Unit 3 Lesson Closer

Option 1: Discuss the recent evidence of mitochondrial DNA linking the origin of man to one woman in Africa. Discuss Dr. Spencer Wells's unique studies collecting blood samples from people of many cultures to look for specific mutations on the Y chromosome that helps him to conclude that all humans alive today are descended from a single African man.

Option 2: Complete the lab, Why Whales Don't Have Legs. This inquiry lab allows students to design a quantitative experiment to show how natural selection favors organisms that are energetically efficient.

Purpose: In this lab, students will propose a reasonable hypothesis as to why whales do not have legs. Students will then design and carry out an experiment to effectively test your hypothesis.

Rules: Students may work in partners *only*! Students many not talk to other groups or ask any questions to the teacher. Students have one double period to design and carry out your experiment. If the data do not make sense, the procedure should be adjusted. Please follow all parts to making a good lab experiment, including controlling variables and multiple trials.

Materials

Container that holds water (bucket)
Two latex gloves
Two small ziplock bags
Hot water (from the sink)
Cold water (from the sink)
Ice
Two thermometers (Celsius please)
The complete lab can be found at http://www.indiana.edu/~ensiweb/lessons/wh.legs.html.

Suggestions for Presenting the Material

- Have interactive discussion of Darwin and natural selection using excerpts from *On the Origin of Species* by Charles Darwin.
- Present PowerPoint lecture notes in outline form.
- Have students create posters showing the endosymbiotic theory, examples of natural selection such as the evolution of the horse, and cladograms. Have students make a pictorial timeline of earth's history.
- Play the flyswatter game with vocabulary. On the board, write up 40 vocabulary words in random order. Divide the class into two teams. Have one team member from each team come to the board. Read a definition, and the first student to hit the correct word (with a flyswatter) earns a point for their team. Repeat and continue with all the members of the group.

Common Student Misconceptions

- Natural selection must include a heritable trait, a selective pressure, differential reproduction and success, and variation in a population.
- Populations only evolve not individuals.
- Evolution has NO GOAL.
- Natural selection and sexual selection balance: The idea that an organisms' display of secondary sexual characteristic cannot decrease chance of survival.
- For Hardy-Weinberg:
    - P^2 = AA
    - 2PQ = Aa
    - Q^2 = aa
    - P = all A in the population from both AA and Aa
    - Q = all a in the population from only aa

Classroom Discussion and Activities

- Film presentations with discussion
    - PBS Evolution series: http://www.pbs.org/wgbh/evolution/
    - *The Simpsons* YouTube clip "Homer Evolution": http://www.youtube.com/watch?v=faRlFsYmkeY
    - *NOVA* evolution site: http://www.pbs.org/wgbh/nova/evolution/
    - *The Triumph of Life* PBS Home Video*: The 4 Billion Year War*
    - 100 Greatest Discoveries: The Origin of Life: Evolution: K-t asteroid and extinction of the dinosaurs; Stanley Miller's experiment; Ballard's Hydrothermal vent explorations PBS: The *Dinosaur Series*
- **Abstract:** Wong, Kate. "Lucy's Baby," *Scientific American*, December 2006. Students will read the above article and write an abstract.
- **Abstract:** Mayell, Hilary. Geneticists Searches for DNA of "Adam," the First Human. *National Geographic News for The National Geographic Chanel.* http://news.nationalgeographic.com/news/2005/06/0624_050624_spencerwells.html
- Jigsaw readings from Science Daily. Have each group read an article on evolution and share with the class what they learned. http://www.sciencedaily.com/news/fossils_ruins/evolution/
- Design an experiment to determine whether the wings of birds, bats, and insects and their ability to fly are the products of convergent evolution. (Russell/Wolf/Hertz/Starr; *Biology: The Dynamic Science*. 1st edition, p. 490.)

- College Board Big Idea 1 labs: BLAST Lab, Hardy-Weinberg Lab, and Artificial Selection Lab Hardy-Weinberg Problems: http://www.biologycorner.com/worksheets/hardy_weinberg2.html
- Hardy-Weinberg lab with Teddy Grahams: http://www.nku.edu/~bowlingb2/HardyWeinbergTeddyGrahams.pdf

Case Studies

The following case studies are from the National Center for Case Study in Teaching Science (http://sciencecases.lib.buffalo.edu/cs/collection/):

- Darwin's Finches and Natural Selection
  - A clicker case where students learn about natural selection through data about population of finches in the Galápagos Islands.
- A Tale of Three Lice
  - A clicker case that explores the questions of how hominins lost their body hair.
- Cross-Dressing Salmon
  - A clicker case about female mimicry in spawning salmon. It addresses one of the most common misconceptions that students have about natural selection, namely, that only the "strong" survive and reproduce.

AP® Practice Essays

1. There are a number of reasons and ways that populations evolve. Describe the mechanisms of micro- and macroevolution by
   (a) comparing the level of occurrence for each.
   (b) explaining at least three different mechanisms for each.
   (c) giving a specific example for each mechanism.
2. Fossils and molecular comparisons among modern organisms inform us about the early history of life including links from prokaryotes to eukaryotes. Describe the links among prokaryotes and eukaryotes including
   (a) unifying processes.
   (b) organelles.
   (c) endomembrane system.
3. The Hardy-Weinberg equation can be used to determine if a population is evolving and what is driving that evolution.
   (a) State and explain three of the five conditions of Hardy-Weinberg equilibrium.
   (b) State the Hardy-Weinberg equation and define its variables.
   (c) In a population of 1,000 moths, 160 show the homozygous recessive trait of white wings. The rest display the dominant trait of gray wings. What is the frequency of the recessive allele? What is the frequency of the dominant allele? How many moths will be heterozygous gray?

Lesson Outline: Unit 5: Plants: Homeostasis and AP® Investigations

Correlates with parts of the 15th edition book, Chapters 22, 28, and 30

- **AP® Biology Big Idea 2:** Biological systems utilize free energy and molecular building blocks to grow, reproduce, and maintain dynamic homeostasis.
- **AP® Biology Big Idea 4:** Biological systems interact, and these systems and their interactions possess complex properties.

Brief chapter summaries

Chapter 22 ("The Land Plants") frames evolution from algae, focusing on the key innovations of vascular tissue, seeds and pollen, and flowers. This chapter flows nicely following the evolution unit, and is an excellent time to begin planting the Wisconsin Fast Plants needed for **AP® Investigation #1: Artificial Selection**, if you haven't already. Some teachers also prefer to have the students do **AP® Investigation #Photosynthesis** at this time rather than during Chapter 6 in order to space out the labs and lab reports that tend to clump together at the beginning of the year if teachers are going through the textbook by chapter.

Chapter 28 ("Plant Nutrition and Transport") describes how vascular tissue helps transport water and dissolved minerals from the roots (transpiration) and phloem from the leaves. This chapter provides excellent background material for **AP® Investigation #11: Transpiration.**

Chapter 30 ("Communication Strategies in Plants") opens students' eyes to the concept that, like humans and other animals, plants have various ways of communicating with each other and with various parts of itself, despite no nervous, endocrine, or immune systems. Phototropism and photoperiodism are also explained in this chapter.

Curricular Note: The AP® Biology curriculum does not require the teaching of any specific plant chapters, though numerous illustrative examples for teaching various Enduring Understandings exist. Additionally, three of the AP® Investigations (#1, #5, and #11) are plant-based labs. Many of the questions on past AP® Biology exams use plants as a base with which to test understanding of other skills and concepts. For that reason, some AP® Biology teachers choose to have a short "plant" unit where they do the plant-based labs and teach students some of the evolution, anatomy, and physiology of plants, especially as it relates to homeostasis.

Note on the organization of this Unit of the IRM: This unit is organized such that all of the relevant AP® Biology Enduring Understanding and Enduring Knowledge goals are at the beginning to help frame the three plant-related labs. Then chapter-specific information for Chapters 22, 28, and 30 are included at the end.

Unit 5 Objectives

- In plants, physiological events involve interactions between environmental stimuli and internal molecular signals.
- Living functions, such as growth and reproduction, require feedback mechanisms to maintain homeostasis.
- Feedback mechanisms help to maintain organisms' internal environments in response to an external environment change.
- Environmental factors can influence an organism's ability to maintain homeostasis and may influence growth.

- Similar feedback mechanisms can reflect a common ancestry among organisms and a difference in feedback mechanisms can reflect an evolutionary divergence.
- In an effort to maintain homeostasis, plants and animals have defense mechanisms that help prevent infection.
- Temporal regulation and coordination affect many homeostatic functions within organisms.
- Regulatory mechanisms within an organism are necessary to maintain homeostasis and often coordinate the timing of such functions.
- Behavior among organisms can affect not only the biological processes within, but can also affect natural selection.

Enduring Understanding 2.C: Organisms use feedback mechanisms to regulate growth and reproduction and to maintain dynamic homeostasis.

Essential Knowledge 2.C.1: Organisms use feedback mechanisms to maintain their internal environments and respond to external environmental changes.

Essential Knowledge 2.C.2: Organisms respond to changes in their external environments.

Enduring Understanding 2.D: Growth and dynamic homeostasis of a biological system are influenced by changes in the system's environment.

Essential Knowledge 2.D.2: Homeostatic mechanisms reflect both common ancestry and divergence due to adaptation in different environments.

Enduring Understanding 2.E: Many biological processes involved in growth, reproduction, and dynamic homeostasis include temporal regulation and coordination.

Essential Knowledge 2.E.2: Timing and coordination of physiological events are regulated by multiple mechanisms.

Warm Up Questions

1. Describe homeostasis and explain why all organisms have to maintain homeostasis to ensure life.
2. Write a hypothesis you may have if you observe a plant growing horizontally toward a light source.
3. List at least three ways you have observed an organism change in response to an external stimulus.
4. What is common ancestry and divergence?
5. Why do you think the internal structure of a snake differs from that of a human?
6. Name three differences and similarities you would expect to see in a plant that lives in the desert versus a plant that lives in the rainforest.

CHAPTER 22: THE LAND PLANTS. Focus on 22.1–22.3

Chapter 22 Lesson Opener

Collect reproductive structures of gymnosperms (use photos, drawings, or models to represent those taxonomic divisions for which structures are unavailable). Compare them to the reproductive structures of angiosperms (students can dissect flowers).

Chapter 22 Suggestions for Presenting the Material

- When students think of plants, almost all of the examples that come to mind will be angiosperms. Counter this tendency by asking for examples of plants that do *not* have flowers.
- Spend time discussing the alternation of generations. This is a difficult concept for students. Try to compare it to the life cycle of an animal.
- Give the students flash cards with pictures of a plant and other relevant traits. Have the students try and classify plants based on their characteristics.
- Review the evolutionary trend from gametophyte dominance to sporophyte dominance. If students still find this confusing, they will have trouble comparing one plant life cycle to another. Highlight evolutionary hallmarks such as development of vascular tissue, dominant sporophyte, heterospory, nonmotile gametes, and seeds that distinguish simple plants from those that are more complex.
- Create a matrix that you and the students fill out to help clarify the evolutionary relationships in the plant kingdom based on several main traits: dominant generation, vascular tissue, roots, leaves, dispersal, reproduction, and size.
- Because of their natural geographic locations, many of the plants mentioned in the earlier sections of this chapter are not familiar to students. Transparencies of representative species, even made from photos in the book, will help remove the abstract quality of a ginkgo or cycad. When you can, however, bring an actual sample or leaf from the tree, it will spark more interest.
- Emphasize that even though this chapter deals with land plants, many are still reliant on water for reproductive purposes. Be sure to explain this in detail.
- Discuss the massive deforestation in British Columbia and other areas of the world. What is being done at the present time to reclaim a natural environment?

Chapter 22 Classroom and Laboratory Enrichment

- **AP® Biology Investigation #5: Photosynthesis**
- **AP® Biology Investigation #1: Artificial Selection**
- Look at fossils of ancient lycophytes, horsetails, ferns, and gymnosperms. Discuss how changing climates influenced the geographic distributions and sizes of these plants.
- Prepare a small display of portions of bryophytes, lycophytes, horsetails, ferns, and gymnosperms from local areas where plant collection is allowed. Prepared slides and live materials ordered from biological supply houses can supplement your collection.
- Go to a location where all of the representatives of the plant kingdom can be observed. Have students keep a journal of where they find each different type of plant, and then have them make inferences about what the plant's location says about the plant's characteristics and requirements.
- Research an area with drastic weather conditions, like Mt. Washington in New Hampshire. What plants are durable enough to withstand the drastic conditions near the top of the mountain? What adaptations allow them to survive in this unfavorable environment?
- The chapter mentions coevolution of flowers and pollinators. Have the students study different pollination syndromes. Have them describe what a flower would have to look like to attract a certain type of pollinator. Show some photos of what common flowers would look like through the eyes of common pollinators. Often there is a pattern that is not evident to the human eye.
- Obtain a set of slides that will provide a survey of the plant world, including algae. Observe some of the components of plants microscopically, such as pollen grains and xylem.
- Which type of cone is most familiar to your students—the male or female one? Why?

Chapter 22 Classroom Discussion Ideas

- What features had to evolve in plants to enable survival on land?
- What adaptations were required for plants to come onto land?
- What adaptations led to the dominance of early vascular plants like ferns? What were key adaptations that led to gymnosperm and then angiosperm domination?
- What are some of the economic and ecological consequences of deforestation?
- What are some of the important adaptations that helped plants survive on land and why were they important?
- Why is pollen an important adaptation to living on land?
- How does the shape of a flower influence its potential pollinators?
- How is honeybee colony collapse syndrome impeding the pollination of some plants?
- Some monks consider ginkgo trees to have a religious significance. How has this helped the species survive? Why do we often find ginkgo trees in intercity habitats?
- What is the role of guard cells in the opening and closing of stomata?
- How are the "leaves" of conifers different from that of the leaves of deciduous trees? How do these differences help the conifers adapt to their environment?
- What major structural difference separates the bryophytes from the ferns?
- What are some differences and similarities between a pine cone and a fruit?
- Why is a gymnosperm considered to have a naked seed?
- Which type of pollination is more efficient—wind pollination, as seen among conifers, or insect pollination, as seen among some of the angiosperms?
- Discuss the various kinds of coevolutionary adaptations seen among plants and their pollinators.
- What adaptive advantage(s) might gymnosperms have had over the other plants living at that time?
- What are several factors that have led to the dominance of angiosperms?

Chapter 22 Possible Responses to *Critical Thinking* Questions

1. The note would have to indicate that there is no pollen source in fern species. Instead, the spores that develop from the "seeds" underneath the fronds develop into the gametophyte generation. When this gametophyte is formed, it has two distinct regions where eggs and sperm are produced. The addition of water allows the sperm to swim to the eggs, and fertilization can occur. In this case, there are no true seeds and the sperm are flagellated, unlike pollen that you would find in gymnosperms and angiosperms.
2. The diploid stage may be more advantageous to genetic diversity. This can be seen in the example of a slightly disadvantageous mutation of a recessive allele that becomes advantageous in the future. Because the allele is disadvantageous, it would be selected against in the population. However, because the allele is recessive, it is only expressed when paired with another recessive allele. If paired with a dominant allele, the recessive allele will have no effect. Because some of the recessive alleles will end up in heterozygous plants, the recessive alleles will remain in the population until a time when they might become more advantageous.
3. Pollen in angiosperms travels via several methods. One important mode of pollen transportation is by the action of specific pollinators. The flowers of angiosperms attract pollinators by looking and smelling attractive. Certain specific signals tend to attract pollinators of the same species, so pollen from the flower of one species is apt to travel to another flower of the same species. This allows for little wasted pollen. Also, some seeds travel through the digestive system of birds or animals, so that the seed coat is broken down and deposited with its own fertilizer. The major method of pollen dispersal for conifers, such as the pine tree, is wind. This method is highly inefficient, because much of the pollen does not contact another tree of the same species.

4. Since bryophytes, such as mosses, absorb most of their water from the air, they also have the secondary effect of obtaining harmful gases from the air. Gaseous pollutants, such as sulfur dioxide, easily dissolve in water. Because of this unique characteristic of mosses obtaining water (along with dissolved pollutants) directly from the air, scientists often examine mosses for deleterious signs of air pollution.
5. Mosses do not have lignin to give them support and they do not have xylem to transport water. With no vascular system, there is no organized way to move water around the plant. A moss is somewhat like a sponge, because it just soaks up water from the air. Therefore, a moss cannot become large without a system of ductwork to transport water and nutrients around the plant. A tall plant would have an advantage, because it has easy access to sunlight. Lower plants have to absorb whatever sunlight passes through the upper canopy of trees.

Chapter 22 Possible Responses to *Data Analysis Activities* Questions

It is hypothesized that insects may assist in the fertilization in mosses.

1. In mosses, the sporophyte is the result of zygote formation that can only occur after fertilization.
2. The mosses had to be 0 cm apart for a sporophyte to form in the absence of mites or springtails.
3. This study showed that the only time fertilization occurred when mosses were at a distance great than 0 cm was if it was mediated by insects.

CHAPTER 28: PLANT NUTRITION AND TRANSPORT. Focus on 28.3–28.6

Chapter 28 Lesson Opener

- Do the first part of **AP® Investigation #11:** Have students collect leaves from campus and/or bring in some from home in order to determine the stomatal location and density using clear nail polish, clear tape, and a light microscope.
- Demonstrate how species adapt to their surroundings by discussing the number, size, location, and distribution patterns of stomata in leaves of different species. Include some unusual examples, such as aquatic plants with stomata on upper leaf epidermis, conifers with sunken stomata, and plants with pubescent leaves.

Chapter 28 Suggestions for Presenting the Material

- Students should be able to make an easy transition to this material if the role of function in determining plant morphology was stressed in presenting the previous chapters. Emphasize that plants are supported by a "skeleton" formed by a continuous column of water inside sturdy xylem tubes.
- Students have a difficult time understanding how water moves. Review some of the terms learned earlier that relate to water movement, such as osmosis and turgor pressure. Simple demonstrations, such as using a wet paper towel versus a dry paper towel to absorb a puddle of water, can be used to demonstrate that water moves toward decreasing water potential.
- To help students understand the large surface area of a plant's root system, provide data on the surface areas of some typical plant root systems. Ask them to guess the ratio of *shoot* surface area to *root* surface area of a typical plant.
- This chapter provides many good opportunities to discuss the selective role of the environment in shaping such features as stomata and root systems.

Chapter 28 Classroom and Laboratory Enrichment

- **AP® Investigation #11: Transpiration**
- Demonstrate the abundance and fragile nature of root hairs. Germinate radish seedlings in a Petri dish lined with paper towels. Pass the dish around the classroom and allow each student to take a seedling and examine the root hairs. What happens to the root hairs minutes after the seedling is removed from the dish? Why?
- Set up demonstrations of root pressure or transpiration in lecture or lab. For example, grow some corn seedlings for 1–2 weeks. Cut the plant, leaving only 2–3 cm of the stem. The root pressure will cause a drop of water to form at the cut.
- Show slides of chloroplasts containing starch grains. Where did the starch come from? Would you be more likely to see such grains in the morning or in the afternoon? What will happen to the starch grains?
- In lab, provide prepared microscope slides of the undersides of plant species from different environments. In lecture, show slides or diagrams of the lower leaf epidermis.
- Grow plants hydroponically and place different concentrations of fertilizer in the water. How much fertilizer is too much? What happens to the plant when too much fertilizer is added?
- Place a stalk of celery in water with food coloring. Over time, slice cross sections from the stem and observe the plant's vascular tissue.

Chapter 28 Classroom Discussion Ideas

- What happens to plants after they have accumulated toxic compounds?
- Genetic manipulation of plants to make more species with the genes required for bioremediation sounds like a good thing. Could it ever be a detriment?
- At what kinds of sites might bioremediation be required? Mines? What types of industries?
- How does the cost of bioremediation compare to the cost of mechanical removal of contaminated sites? This information is readily available on the Internet.
- How do the tracheids and vessel members found in xylem conduct water even though they are dead at maturity? In contrast, why must phloem cells remain alive to perform their function?
- Desert plants must balance the need for carbon dioxide against the threat of desiccation. What are some adaptations of desert plants that allow them to open their stomata often enough to get the carbon dioxide sufficient for photosynthesis? Discuss how alternative photosynthetic pathways, such as C4 and CAM photosynthesis, have evolved in response to environmental pressures.
- What are the invasive species of plants that are a problem where you live? Track the origin of these species and discuss how they traveled to the area.
- Research the history of your state flower and/or tree. Why were they chosen?
- What happens to transpiration rates on hot days? Dry days? Humid days? Breezy days?
- Ask students who have raised tomatoes or other garden plants if they have ever observed "midday wilt," a phenomenon in which even well-watered plants temporarily wilt during the late afternoon. Ask them why this happens. (Midday wilt occurs when transpiration exceeds the rate of water uptake.)
- In what ways are animals dependent on plants for survival?
- How do plants combat insect pests in nature? How do we help plants resist insect attack?

Chapter 28 Possible Responses to *Critical Thinking* Questions

1. Insufficient nitrogen in a plant will have a negative impact on the synthesis of amino acids and the proteins from which they are assembled. This will, in turn, affect the structure of the plant, which is incorporating proteins, plus it will curtail the synthesis of enzymes needed for metabolic activities. Lack of nitrogen will also adversely affect the synthesis of nucleic acids needed for information storage (DNA) and decoding (RNA).
2. Water is important for almost all cellular functions. When water is abundant, plants store water in vacuoles, which increases the turgor pressure of cells, causing them to be plump. In addition, when water is in abundance the stomata open during the day, allowing photosynthesis to take place, but close at night in order to conserve water. During your vacation as the soil became dry, the root cells released abscisic acid, which in turn released nitric oxide. This caused an increase in calcium ions in the guard cells, which pumped potassium and chloride ions, malate, and water out of the guard cells, causing them to collapse. The collapsed guard cells closed the stomata in an effort to conserve the plant's water. As the plant remained in the dry soil, the only water available to maintain cellular functions was in the plant's vacuoles. As water was removed from the vacuoles, the turgor pressure decreased and the cells became less plump, causing the plant to wilt. Plants can survive wilting to a degree; however, at some point the plant will have lost too much water and will not recover even if abundant water is applied.
3. When moving a plant from one location to another, it is a good idea to take as much soil as practicable with the root mass. This will ensure that the mycorrhizae are substantially intact and can continue functioning in the new location, and that the fragile root hairs so essential for water/mineral absorption are not stripped off.
4. Stomata allow carbon dioxide, used during photosynthesis to make sugar, to flow into the plant. The byproduct of photosynthesis is oxygen, which must be released from the plant. Stomata also allow the plant to regulate water loss. If the stomata remain open, too much water could be lost and the plant will wilt and die. If the stomata remain closed, no gas exchange will occur and no photosynthesis will take place.
5. The three micrographs are all from the vascular tissue known as xylem. Micrograph A is a close-up of a perforation plate located at the ends of the vessel members. Micrograph B shows three vessel members stacked on top of each other; note that at the junction of each vessel member is a perforation plate. Micrograph C shows tracheids. Notice that tracheids are thinner and more pointed than vessel members. Tracheids are connected by pits.

Chapter 28 Possible Responses to *Data Analysis Activities* Questions

1. Six. Five were planted and one was not planted in the soil (unplanted).
2. The unplanted transgenic control shows very little, or no, uptake in the first 3 days of the experiment; in that experiment, the TCE concentration remains at 15,000 $\mu g/m^3$ for the first 3 days. The planted transgenics remove TCE from the air the fastest. This is seen as a decrease of TCE from ~15,000 $\mu g/m^3$ to less than 10,000 $\mu g/m^3$ in less than 2 days.
3. The air around the transgenic plants contained less than 5,000 $\mu g/m^3$. The air around the vector control plants contained more than 15,000 $\mu g/m^3$.
4. One explanation is that the transgenic plants are removing the TCE from the air and incorporating it into the plant tissue. Alternatively, the transgenic plants could be releasing a volatile compound that converts the TCE to another compound. If the air was measured for TCE content and the presence of other compounds, the researchers might be able to distinguish between these two possibilities. Alternatively, they could harvest the plant material from the plants and determine if the transgenic plants contained more TCE incorporated into their plant tissue.

CHAPTER 30: COMMUNICATION STRATEGIES IN PLANTS

Chapter 30 Lesson Opener

- Introduce the term *hormone* and discuss the need for growth-regulating hormones. There are many lab experiments designed to investigate the role of hormones in plant growth and development; many of them are available from biological supply houses in the form of kits. These can be prepared ahead of time and used as demonstrations or performed by the students. These kits can be used to examine the effects of auxins, gibberellins, and cytokinins on plant growth and development.

Chapter 30 Suggestions for Presenting the Material

- You can ask students to think about the common gardening practices presented next that involve concepts presented in this chapter to give them familiar reference for the more detailed information in the sections. Ask them to explain why (and how) the following strategies work:
  1. Keeping fruit from ripening: Fruit should be kept in a space with a constant flow of oxygen.
  2. Growing vegetables from seeds: Seeds need to be planted in soil that can drain well and they must be watered thoroughly.
  3. Pruning: Cutting the tip off the shoot of a plant will make the plant grow bushier.
  4. Forcing bulbs: Bulbs must be kept cool for at least 12 weeks before they will bloom inside.
  5. Growing a tree from a seed: Seeds should be dried until the seed coat cracks before they are planted.
- Discuss with your students once again how growth in plants differs from growth in animals.

Chapter 30 Classroom and Laboratory Enrichment

- Design and implement experiments involving seed germination. Vary conditions such as temperature or light, and examine the effects on germination rate.
- How does complete darkness affect seedling growth? Compare the lengths and weights of bean or pea seedlings raised in light with those raised in complete darkness.
- Investigate the role of the intact shoot tip in inhibiting lateral bud development in *Coleus*. What happens to lateral buds if the shoot tip is removed? Design an experiment in which applications of IAA dissolved in a paste of lanolin mimic the effects of the intact shoot tip.
- Discover the effects of different wavelengths of light on phototropism. Different colored films can be used to filter out certain wavelengths of light. For example, a red theatrical filter will absorb blue and transmit red.
- Dwarf varieties of plants can be purchased from biological supply companies. The dwarf phenotype can be rescued (in some varieties) by applying gibberellin. You could determine if the amount of gibberellin applied correlates to the amount of rescue.
- Examine the effects of ethylene on fruit ripening or abscission. For example, place bananas in a bag with or without apples and observe the difference in ripening. How does "one bad apple spoil the bunch"?

Chapter 30 Classroom Discussion Ideas

- Why do many fruits change from green to red, yellow, or orange as they ripen?
- The first transgenic plants sold for human consumption were tomatoes that had been modified such that no ethylene was produced. What are the benefits of these tomatoes? Potential risks?

- How do hormone production and activity differ in plants and animals? What are some similarities shared by plants and animals? What are some differences?
- What is "pinching back"? Why does it make plants bushier?
- What is "bolting"? Ask any of your students who have raised spinach or lettuce if they are familiar with this term. What can gardeners do to inhibit bolting?
- How could a sap-feeding insect actually stimulate plant growth? (Answer: The plant works harder to replace lost sap.)
- How could a foliage-feeding insect increase growth of understory plants? (Answer: increased light penetration)
- How could irrigation of cotton plants increase insect damage? (Answer: More green leaves support more larvae.)
- Discuss the role plant movement plays in flowering plants. What is the adaptive purpose(s) of this movement?

Chapter 30 Possible Responses to *Critical Thinking* Questions

1. During the normal germination process, abscisic acid (ABA) levels must first decline. Since the presence of ABA represses expression of genes that are involved in growth, when ABA levels decline, these genes are now expressed. One of the genes expressed as ABA levels decline is the gene that encodes NADPH, an enzyme that produces hydrogen peroxide in the cell. Hydrogen peroxide enhances the expression of genes involved in gibberellin synthesis. Since gibberellin promotes growth, hydrogen peroxide promotes germination and seedling growth by increasing the amount of gibberellin in the seed. Professional gardeners who soak seeds in hydrogen peroxide are promoting gibberellin synthesis.
2. Plants grow tall and spindly when they do not receive enough light, since light inhibits stem elongation. Phytochrome receptors of blue light cause the cessation of straight growth in seedlings when they are exposed to light. Flavoproteins absorb the blue light wavelengths, which results in a redistribution of auxins. Seedlings germinated in darkness are experiencing greater amounts of hormones. These could be varying amounts of gibberellins, ethylenes, and auxins.
3. Excess auxin production in the wall cress plant would result in increased hypocotyl elongation in the light, increased apical dominance, and severe downward leaf growth. Further effects from the excess auxin could include dramatic developmental defects including embryo and emerging leaf symmetry anomalies, premature inflorescence development, altered leaf arrangements along the stem, reduced petal size, abnormal stamens, sterility, and root-growth defects.
4. There are major differences between how plant and animal hormones function, and it is unlikely that a plant hormone would influence growth and development in humans. Human cells do not have the same hormone receptors and signaling pathways that a plant does; therefore, the plant hormone would not be able to initiate a signaling pathway in humans. An exception may be phytoestrogens, which are plant hormones that are chemically similar to estrogens. Phytoestrogens are weaker than human estrogens but they do cause hormonal effects in humans. Some studies have shown that phytoestrogens positively influence human health by protecting against cancer; however, other studies have shown that high levels of phytoestrogens can cause reproductive problems.
5. In the tip of a coleoptile are light-sensing pigments called phototropins. When the phototropins absorb light, auxin efflux carriers accumulate on the side of the plant that receives the least amount of light. This causes more growth on the shaded side of the plant; thus the plant bends toward the light. If you remove or cover the light-sensing tip, as in A and C, the plant cannot detect light and no bending will occur. Both B and D will bend toward the light because the tip is able to sense the light and the efflux carriers can accumulate on the shaded side of the plant.

Chapter 30 Possible Responses to *Data Analysis Activities* Questions

1. Plants attacked by budworms and thrips (HVT) or budworms alone (HV) produced a total of 11 different volatiles in an amount of 13,563 ng/day and 9,423 ng/day, respectively.
2. When plants were exposed to budworms and thrips (HVT), 6,166 ng/day of β-caryophyllene was produced.
3. Nonspecific means that the compound is produced whenever there is damage/threat to the plant, regardless of the threat. In this case, the answer would be β-caryophyllene because it is produced in all cases of damage/threat, but not in the control.
4. Plants attacked by thrips (T) produce β-ocimene, β-caryophyllene, sesquiterpene, and α-farnesene, while plants attacked by budworms (HV) produce these chemicals plus myrcene, linalool, indole, nicotine, β-elemene, α-humulene, and caryophyllene oxide.

UNIT 5 Homework Extension Questions

1. Name two types of plants that have homeostatic mechanisms that would suggest divergent or common ancestry.
2. How does the environment regulate how nutrients are taken in and how waste is expelled?
3. How do adaptations to the environment reflect common ancestry and divergence?
4. Describe what migration is, and indicate if it is a positive behavior, negative behavior, or both.
5. What homeostatic mechanisms do doctors measure when they take vital signs such as temperature, pulse, and blood pressure?
6. What are the effects of too much or too little growth hormone?

Lesson Outline: Unit 6: Organisms

Correlates with Chapters 32, 34, 37, 41, and 42 in the 15th edition

- **AP® Biology Big Idea 2:** Biological systems utilize free energy and molecular building blocks to grow, reproduce, and maintain dynamic homeostasis.
- **AP® Biology Big Idea 4:** Biological systems interact, and these systems and their interactions possess complex properties.

Brief chapter summaries

Chapter 32 ("Neural Control") is an information-packed chapter focusing on how neural cells function to help achieve homeostasis and communication within an organism as well as in response to environmental stimuli. The most relevant sections of Chapter 32 for the AP® Biology curriculum are 32.1–32.7. Some teachers include a few sections of Chapter 33 ("Sensory Perception") to highlight the interaction between the environment and the organism.

Chapter 34 ("Endocrine Control") gives students a concise, thorough treatment of how communication can occur via the bloodstream, not just at the synapse, focusing on various glands, their hormones, and descriptions of disrupted homeostasis. Some teachers bring in parts of Chapter 41 ("Animal Reproductive Systems") to further highlight hormonal regulation.

Chapter 37 ("Immunity") describes how animals resist and fight infection. Comparing this chapter to how plants resist and fight infection is a good idea in the AP® Biology framework.

Chapter 42 ("Animal Development") revisits the role of master genes, differential gene expression, and cell-to-cell communication in the context of how a zygote develops into an adult, especially focusing on humans.

Curricular Note: The AP® Biology curriculum does not require the teaching of any specific body systems, though there are numerous instances where concepts directly related to specific body systems exist. Because of that, and because students tend to enjoy the body systems unit, many teachers choose to teach all or parts of the aforementioned chapters.

Note on the organization of this Unit of the IRM: Similar to the plant unit, this unit is organized such that all of the relevant AP® Biology Enduring Understanding and Enduring Knowledge goals are at the beginning. Then chapter-specific information for the chapters is included at the end.

Unit 6 AP® Biology curriculum framework connections

Enduring Understanding 2.D: Growth and dynamic homeostasis of a biological system are influenced by changes in the system's environment.

Essential Knowledge 2.D.3: Biological systems are affected by disruptions to their dynamic homeostasis.

Essential Knowledge 2.D.4: Plants and animals have a variety of chemical defenses against infections that affect dynamic homeostasis.

Enduring Understanding 2.E: Many biological processes involved in growth, reproduction, and dynamic homeostasis include temporal regulation and coordination.

Essential Knowledge 2.E.1: Timing and coordination of specific events are necessary for the normal development of an organism, and these events are regulated by a variety of mechanisms.

Essential Knowledge 2.E.2: Timing and coordination of physiological events are regulated by multiple mechanisms.

Essential Knowledge 2.E.3: Timing and coordination of behavior are regulated by various mechanisms and are important in natural selection.

Enduring Understanding 3.B: Expression of genetic information involves cellular and molecular mechanisms.

Essential Knowledge 3.B.2: A variety of intercellular and intracellular signal transmissions mediate gene expression.

Enduring Understanding 4.A: Interactions within biological systems lead to complex properties.

Essential Knowledge 4.A.4: Organisms exhibit complex properties due to interactions between their constituent parts.

Unit 6 Warm Up Questions

1. How do cells interact with one another?
2. Why is intercellular communication necessary?
3. Explain the saying, "One rotten apple spoils the whole barrel."
4. What is immunity?
5. What are the effects of desertification and deforestation on an ecosystem?
6. What influences the movement of ions and molecules?
7. Do plants have any defense against predation?
8. How can animals defend themselves against harmful organisms such as bacteria or viruses?
9. Do animal immune systems ever have problems?
10. What is genetic engineering?
11. Describe the process of transcription and translation.
12. Why is cell death important in maintaining homeostasis?
13. Name some environmental factors that begin seed germination.
14. What is a circadian rhythm?
15. What factors influence animals to migrate?
16. What is natural selection?
17. Name three responses plants have to changes in the seasons.
18. Why is it better for survival if species can communicate?
19. What are organ systems made of?
20. How are animal bodies organized?
21. How does the structure of a human digestive tract help with its function?

Unit 6 Lesson Opener

Speedy: (10–15 minutes) Instruct students to record observations they have made about themselves or others concerning their physical condition during extreme heat and cold conditions. List each observation on the board and discuss how each physiological response helps the body maintain homeostasis.

Extensive: (40–50 minutes) Turn half the lights off in the classroom before students enter the room. Give them a short activity, such as a word search with approximately 20 words, and a stopwatch. Allow them to record the amount of time needed to complete the word search. When everyone is finished, instruct the students to observe changes in the eyes of their neighbors when the lights are turned back on. With the lights on, give them time to complete the word search with 20 new words. Instruct them to record their time once again.

Discuss the physiological change that happened to their eyes once the lights were turned on. Discuss how the time required to complete the task changed with the lights on versus with the lights off. Ask students to explain how their physiological adaptation related to the change or lack of change in time to complete the task.

CHAPTER 32: NEURAL CONTROL. Focus on 32.1–32.7

Chapter 32 Lesson Opener

Some students may be familiar with the children's game called "whisper down the lane." This game can be used to illustrate the transmission of a nerve impulse. First, line up a row of students from one side of the room to another. Each student should pass a message along, one by one, through the line until the message gets to the end. This process illustrates the transmission of a nerve impulse when it does not contain a myelin sheath. To mimic the transmission in a myelinated neuron, a student would lean over and give the message to a student several places away. The message will get to the end faster, because the impulse skips major portions. This is similar to the transmission of a nerve impulse when the axon is myelinated.

Chapter 32 Suggestions for Presenting the Material

- To grab the students' attention, start the lecture with any of the ABC videos available from the Multimedia Manager and CengageNOW.
- Students ranging in scientific expertise, from that of beginning freshman to third-year medical student, tend to agree that the nervous system is one of the most difficult to comprehend at any level. Therefore, extra time and thorough explanations are especially needed in this chapter.
- The function of the neuron membrane in permitting passage of Na^+ and K^+ ions is confusing. Initially, you may wish to focus on sodium only, and then expand to the role of potassium.
- The changes in the neuron membrane can be conveniently demonstrated using Figure 32.8 (a–d) and the words *polarized*, *depolarized*, and *repolarized*, which are conventional terms but are not used in this edition of the text. The Multimedia and CengageNOW provide an animation as well.
- The concept of "all-or-nothing events" and "thresholds" can be illustrated by describing the use of a firearm. When the trigger is pulled and reaches the critical point (threshold) at which the hammer is released, the bullet leaves the barrel and travels the expected distance. Of course, the bullet either goes or stays (all or nothing), and the manner in which the trigger is activated (slowly or quickly) should not influence the speed of bullet travel. Keep in mind that, at any given time, a gun is either firing or not firing—just like a neuron is either resting or excited. Appropriately enough, scientists even describe a nerve impulse as "firing"!
- Equally effective in illustrating the "all-or-nothing" concept is the use of a device present in every classroom—light switches.
- Emphasize the temporary nature of the acetylcholine bridge across the synapse by comparing it to a pontoon bridge used by the military to cross small streams and rivers.

- If the students can recite the sequence of structures through which an impulse passes during a *reflex arc*, such as in Figure 32.18, they have a good grasp of the nerve conduction pathways. Add the ion flow across the membranes, and the story is pretty well complete!
- Emphasize the continuity of fiber tracts between the brain and spinal cord. Stress the primary functions of the spinal cord as a reflex center versus the brain as a sense-interpretation and directed-response center.
- The extent to which each instructor requires the students to delve into brain regions and functions will vary. Some instructors may want to select the major brain regions (cerebrum, hypothalamus/pituitary, cerebellum, and medulla) for special emphasis.
- Students generally find it easier to distinguish between the sympathetic and parasympathetic divisions of the autonomic nervous system if they are told that the sympathetic division is involved in mobilizing "fight-or-flight" reaction, while the parasympathetic division produces a general "slowing-down" and "business as usual" response. Use an example of the functioning of each system that students will relate to easily. For example, if you see a white car by the side of the road that you perceive to be a police car, your sympathetic nervous system may respond by speeding up your heart rate. As you get closer and realize that it is simply a disabled car, your parasympathetic system returns your heart rate to normal.

Chapter 32 Classroom and Laboratory Enrichment

- The concept of thresholds and all-or-nothing events can be demonstrated by using dominoes (or for large classes, several audiocassette cases). Line up about 20 dominoes placed on end and spaced about 1 inch apart. Ask a student to gently touch one end domino to begin the progressive fall. Emphasize that the student's touch (threshold stimulus) caused a standing row (polarized) to begin falling (depolarization) at a constant speed (all-or-nothing event). Pose the following question (and demonstrate the answer): Would a greater and faster stimulus cause more rapid falling? To demonstrate *repolarization*, a second student could begin resetting the dominoes even before the falling is complete.
- Permit students to demonstrate the knee-jerk reflex arc by using percussion hammers. It is important to ask the subjects to close their eyes to prevent "cheating."
- Show a film or video of an animation of nerve impulse transmission.
- Exhibit models of neurons and neuroglial cells.
- Provide microscope slides of longitudinal sections and cross-sections of nerves for student viewing.
- Have microscope slides of neurons available for laboratory demonstration. Be sure to point out the long cytoplasmic extensions—dendrites and axons.
- Many laboratories have preserved specimens of the human or other vertebrate brain, which are valuable aids to comprehending the size and arrangement of brain parts.
- Use a dissectible model of the neurons, neuroglial cells, the brain, and brain tumors to illustrate the important concepts.
- Show a film or video on brain function.
- Use a spinal cord/vertebral column cross-sectional model to illustrate the relation between the two structures.

Chapter 32 Classroom Discussion Ideas

- Why are athletes, such as NFL players, particularly susceptible to concussions? What are other risk factors?
- Borrow brain models from an anatomy lab. Can you identify the major portions listed in this chapter? Refresh your memory of the functions of each of these parts.
- Would changes in the way contact sports are played prevent concussions?
- The information on the psychoactive drugs and their deleterious effects on the body may discourage the informed biology student from this type of drug experimentation.
- What would be the result of demyelination of axons, such as that which occurs in multiple sclerosis (MS)? On first thought, you may think the destruction of myelin sheaths would not be that problematic for the patient. After all, there are neurons in the body that are never wrapped in myelin sheaths and they work perfectly. The problem that results from MS is that the destruction of the myelin sheaths leaves behind scarring, which may impede the conduction of the nerve impulse.
- On hearing that salt was not good for him, a freshman college student began a fanatical program to eliminate all sodium chloride from his diet. By cooking his own meals, he was able to eliminate virtually all sodium. What complications could he expect as a result of his rash action?
- If neurons operate under the all-or-nothing principle, how are we able to distinguish soft sounds from loud sounds, or a gentle touch from a crushing blow?
- Why does a physician's tapping of the knee or elbow reveal the status of the nervous system *in general,* not just the condition of those two joints?
- Why does "jumping" conduction afford the best possible conduction speed with the least metabolic effort by the cell?
- Do you see why the neurotransmitter must be removed after a nerve impulse? If not, the neuron would be confused—is that the last impulse or do you want me to contract again?
- Talk about the fight–flight response. Note how it is designed to give the individual a little extra speed or strength in an emergency. Often, we feel this response in a perceived emergency. Ask someone in the class to get up and sing a song or give a speech. Often the heart rate increases due to the fight–flight response. Some individuals with adrenal tumors have a constant feeling as though they were in a dangerous situation. Can you imagine how stressful that would be?
- Do invertebrates, such as the cockroach, feel pain?
- The central nervous system (and closely associated ganglia) houses the cell bodies of neurons. As opposed to the peripheral axons and dendrites, the cell bodies are not regenerated after traumatic injury. What advantages and disadvantages does this structural arrangement pose for humans?
- Explain why elderly people may be unable to remember what they ate for breakfast but can relate the details of a teenage romance.
- Discuss the characteristics of brain disorders such as Parkinson's disease and Alzheimer's disease. Investigate potential causes of these diseases by looking into recent research.
- Read about Phinneas Gage, a survivor of a traumatic brain injury. From observing the changes in his personality after his accident, do you think the limbic system was damaged?

Chapter 32 Possible Responses to *Critical Thinking* Questions

1. PTSD is known to affect the hippocampus and the amygdale, which are both components of the limbic system. The limbic system works with the cerebral cortex, affecting emotion and contributing to memory. Other components of the limbic system and the cerebral cortex may also be disrupted in PTSD. Also, since it is known that the hippocampus and amygdala are affected, other internal homeostatic controls may be out of balance in patients. For instance, anxiety, depression, and social phobias are also common in PTSD victims.
2. All too often we think of babies as miniature adults, but they are not. The undeveloped blood–brain barrier is one good example. The adult body can process perfectly normal hormones, amino acids, ions, and the like, even if their concentrations vary somewhat to the high side. Even alcohol, caffeine, and nicotine can be detoxified. But to the susceptible fetus and newborn, any of these substances could enter the brain in concentrations capable of causing damage. Thus, in addition to paying careful attention to her newborn's diet, a conscientious nursing mother would limit her indulgences for the health of her newborn baby. Both parents should continue to be careful in the foods they provide to their baby as he or she develops.

Chapter 32 Possible Responses to *Data Analysis Activities* Questions

Studies have been performed to see the effect on the offspring of pregnant rats that were exposed to MDMA (the active ingredient in Ecstasy).

1. During the first 5 minutes, the rats that were prenatally exposed to saline moved around more than those exposed to MDMA.
2. The rats exposed to MDMA prenatally showed 76 photobeam breaks during their second 5 minutes in a new cage.
3. The rats exposed to MDMA moved around the most in the last minutes, when compared to rats exposed only to saline.
4. This study appears to show some differences in the brains of those rats exposed prenatally to MDMA and those exposed to just saline.

CHAPTER 34: ENDOCRINE CONTROL

Chapter 34 Lesson Opener

An excellent case study designed to teach much of the endocrine system in a real-world manner is *Mary Keeper's Aching Head.* http://sciencecases.lib.buffalo.edu/cs/files/aching_head.pdf

Chapter 34 Suggestions for Presenting the Material

- The "core" of this chapter is the section describing the various endocrine glands, their secretions, and functions; Figure 34.2 and its animation are excellent aids.
- Other important topics include signaling molecules and feedback loops.
- When discussing signaling molecules (Animation/Figure 34.3), emphasize the chemical nature of steroids (lipid-bilayer soluble) versus proteins (not lipid soluble) and the need for a second messenger, cyclic AMP. Playing this animation will go a long way to providing visual context for the array of new terms.
- Because the chapter contains excellent figures and tables, these can be used to great advantage. You can request for students to bring their texts to class and invite them to follow along. You can also prepare your own overhead transparencies or slides of the figures and tables to help everyone keep up the pace.

- This chapter presents a maddening array of new words—hormone names that are long and unfamiliar. As an aid to learning, subdivide the name and give the literal meaning of each portion, for example, adreno (adrenals), cortico (cortex), and tropic (stimulate).
- Emphasize the necessity of learning both the hormone name and its abbreviation, which is often more commonly used than the name itself.
- Take this opportunity to point out antagonistic hormone pairs: calcitonin/parathyroid hormone and insulin/glucagon.
- Notice that even though the gonadal hormones are included in Table 34.3, they are not discussed thoroughly until later chapters.
- Emphasize that the posterior pituitary gland does not synthesize the hormones it secretes.
- Point out that some organs function as both endocrine and exocrine glands.
- Discuss the impact of diabetes on humans and the role our diet and lifestyle play in the perpetuation of this disease.
- The students may be amazed that males and females produce some of the same reproductive hormones. Note that the hormone names were based on their functions in females.

Chapter 34 Classroom and Laboratory Enrichment

- Human nature is such that students are very interested in the abnormalities that hyper- and hyposecretion of human hormones cause. You can stimulate interest in the overall area of hormone control by showing slides or transparencies of the physical manifestations of such imbalances.
- If a member of the class is willing to share his or her experiences as a diabetic, arrange for such a presentation before class begins and allow time for questions.
- Seek evidence of a class member who has experienced or witnessed an epinephrine-mediated "emergency response." Ask him or her to report. Is anyone in the class required to carry an "EpiPen"?
- Survey local grocery stores to determine the relative stocks of iodized and noniodized salt. Are there implications for the unwary consumer?
- Use a dissectible mannequin or a dissected fetal pig to show the locations of the various endocrine glands.

Chapter 34 Classroom Discussion Ideas

- If atrazine is found to be harmful to amphibians after further ongoing research, how might farmers who have depended on atrazine react if it is proposed that it be removed from the market?
- As humans of today are dying of cancer and heart disease, what were the major diseases of 100 years ago?
- Discuss the impact hormones in our environment and food are having on changing the onset of puberty and sex traits in animals.
- Look into the use and eventual ban of other chemicals that have deleterious effects on animals, such as DDT and PCBs. How long did it take for scientists to convince lawmakers to restrict their use?
- Why is the pituitary often called the master gland? Do you think that this title is well-deserved since the hypothalamus directs many pituitary activities?
- Be sure to stress to the students that both the nervous system and the endocrine system control activities of the body. The difference is in the way that the control is administered—the nervous system does so via nerve impulses and the endocrine system utilizes hormones.

- Once a student learns the function of a hormone, it may be easy for him or her to visualize situations where hypo- or hypersecretion takes place. The secretion of human growth hormone is an ideal example. Hyposecretion results in pituitary dwarfism, and hypersecretion is called gigantism. What is the treatment for pituitary dwarfism?
- Mention to the students that if a patient takes too many medicinal steroids, he may exhibit symptoms similar to Cushing's syndrome. This makes sense since the adrenal hormones involved in Cushing's syndrome are categorized as steroids.
- Beverage alcohol inhibits the action of ADH. How is this unseen physiological event evidenced during a night of bar-hopping? Can you think of other substances that inhibit the action of ADH?
- Do hormones occur only in vertebrates? What is the function of arthropod hormones?
- Discuss the research regarding pancreatic cell transplants.
- Why does insulin have to be administered by injection rather than orally?
- Using knowledge gained in a freshman biology class, an athlete decided he might be able to raise his blood sugar quickly by injecting glucagon. This attempt is doomed for what reasons?
- What is the possible connection between the pineal gland and puberty?
- What is the connection between the pineal gland and sunlight?
- Some hormones seem to be doing another's duties, for example: sex hormones from the adrenals, blood sugar control by epinephrine, and thyroxine regulation of growth. Why is this so?
- What are anabolic steroids? Why do some athletes use them? What are the dangers associated with their use?
- Oxytocin is commonly used to induce labor. How does it work?
- Why do certain hypoglycemics, who regularly ingest excessive amounts of sugar, frequently develop diabetes later in life?
- Untreated diabetes mellitus victims tend to be very thirsty and yet produce large volumes of urine. Why is this so?
- Women may experience some mild uterine contractions while nursing their babies. Why is this beneficial to a woman's post-pregnancy health?

Chapter 34 Possible Responses to *Critical Thinking* Questions

1. It is hypothesized that totally blind women are less likely to develop breast cancer due to the hormone melatonin. In many Westernized and industrial nations, increased exposure to artificial light suppresses melatonin secretion. This effect is reduced in individuals who are blind.
2. If a diabetic injects too much insulin, it may lower the blood glucose to a dangerously low level, because the brain requires glucose as fuel to perform basic functions. If glucagon is administered, it will raise the blood sugar again—hopefully to a normal level. This example illustrates how important it is that the proper amount of insulin be administered to diabetics.

Chapter 34 Possible Responses to *Data Analysis Activities* Questions

In previous studies, agricultural chemicals were known to decrease the sperm count in some animals. These data were collected to see if there is a similar effect on humans.

1. The highest sperm count was in NYC and the lowest was in Columbia, Missouri.
2. The highest sperm motility was in NYC and the lowest was in Columbia, Missouri.
3. In Missouri, young age and low rates of STDs seem to correlate with low sperm counts and percent motility, while older men with greater rates of STDs seem to correlate with high sperm counts and percent motility in NYC; both groups had similar rates of smokers. In Los Angeles, young age and

the greatest proportion of smokers seem to correlate with high—but not the highest—sperm counts and high percent motility. Minneapolis, with the lowest rates of smokers and a high rate of STDs, had men with high sperm counts and motilities. While it is possible that smoking, age, and STDs could explain the variations in sperm count and motility in these regions, it is impossible to tell from these data, given the many variables and lack of control groups.

4. These studies suggest that living near farmlands can result in lower sperm counts.

CHAPTER 37: IMMUNITY

Chapter 37 Lesson Opener

Show Episode 1, Parts 3 and 4 ("Body Snatchers") of the series called Body Story. As students later read the textbook, have them refer back to the movie to articulate which details the movie represented and which ones they didn't as a way to solidify the material.

Episode 1 Part 3: https://www.youtube.com/watch?v=D2Uuc76DvlI
Episode 1 Part 4: https://www.youtube.com/watch?v=soIDiqxtTjM

Chapter 37 Suggestions for Presenting the Material

- A quick glance at the figures in this chapter should convince anyone that the subject of body defense mechanisms is a difficult one both to teach and to learn. However, with patience and explicit use of those same figures, the topics can be mastered.
- The understanding of defense against foreign organisms by the human body is complicated by the fact that so many mechanisms and factors are operating at the same time. Remind students to use summary tables and animations provided in CengageNOW to help with this information.
- You may wish to compare the "lines of defense" mentioned in the chapter to the military defense of a country. There are many analogies such as "general barriers" (moats aren't very useful anymore) and "scorched earth" (thank you, General Sherman). Nowadays, missiles (third line?) are very specific and fairly accurate.
- Students sometimes have difficulty distinguishing "antibody" from "antigen." This may help: Antigen is short for ***anti***body ***gen***erator. An antigen is something that encourages the body to make an antibody against it.
- The function of the thymus gland is expanded here in more detail from its introduction in the discussion of the endocrine system; remind students of the interconnection between endocrine control and immunity in addition to circulation and neural control.
- The topics of *immunization* and *immune diseases* are always of interest to students and should be given sufficient time to allow for student discussion.
- Remind the students that they have all had experiences with lymph nodes. When a patient complains of "swollen glands," he or she is actually mentioning the fact that lymph nodes in the neck area are undergoing an immune battle due to a foreign invader.
- Remind the students that phagocytosis is a nonspecific process—the macrophage engulfs something that it determines is nonself. On the other hand, antibody production is a specific approach. The B-cells make antibodies to a specific enemy, such as the measles virus.

Chapter 37 Classroom and Laboratory Enrichment

- The dramatization of the suffering and death of a victim of AIDS (or a similar disease) will serve as an attention getter for this topic.
- Tracking down the cause of an annoying allergy can involve some real detective work. Survey the class for such an experience, and ask for a brief oral report if the person is willing to share his or her experience.
- Ask an allergist to address the class, if available. How can an individual decrease his or her reaction to an allergen via a series of injections? Are there some allergies that are so strong and capable of eliciting an anaphylactic reaction, they cannot be lessened by allergy shots? Are some allergies life-threatening?
- With the assistance of a microbiology student, prepare Petri dishes onto which smears from the human mouth, nose, hands, head, as well as commonly touched surfaces are made. Identify the microorganisms present in each location.
- Does an insect's body carry many germs? Attempt to answer this by letting different insects including a cockroach, house fly, and cricket crawl over the surface of an agar-filled Petri dish.
- Interview a hospital epidemiologist to find out how a hospital environment fights the spread of infections, particularly MRSA.
- In literature, milkmaids are always described as pretty. This is probably because they were immune to smallpox because of their exposure to cowpox. Therefore, they would not have the disfiguring scars of a smallpox victim.

Chapter 37 Classroom Discussion Ideas

- How is AIDS perceived as a disease in differing countries of the world (homosexual, heterosexual, prostitutes, etc.)?
- What could be done to stem the epidemic of AIDS in African countries?
- Why hasn't anyone found a vaccine for AIDS yet?
- Why was the AIDS epidemic ignored for such a long time? Was the blood bank community slow in responding to protect the blood supply?
- Since advent of the AIDS epidemic, universal precautions have been instituted in medical settings. Under these conditions, everyone is considered dangerous, so you would not expose yourself to the blood or body fluids of any patient.
- Every year a small number of children die from diseases that develop as a result of vaccines received to protect them. It seems to be an inherent hazard associated with mass preventative inoculation. Is it worth the risk? Can you debate both sides of the issue?
- Thomas Malthus proposed three "grim reapers" that would restrain human population growth. One of these was "pestilence," or disease. How effective is disease as a population-limiting factor in the developed countries versus the underdeveloped countries?
- If there are so many infectious people as patients in hospitals, why aren't doctors and nurses continuously ill?
- What happens when parents refuse to inoculate their babies as mandated by various state and federal agencies? What are the inoculations for?
- Cancer of the thymus gland is called a thymoma and is not common. If one of the treatments of this cancer is to remove the thymus gland, what might happen to a 30-year-old patient? What if the patient is 2 years old?

- Fever and inflammation are natural responses of our immune system. Is it wise to take over-the-counter medications to reduce these responses?
- Make sure the students understand the difference between the functions of T-cells and B-cells. B-cells make an antibody after the T-cells present the foreign object to them.

Chapter 37 Possible Responses to *Critical Thinking* Questions

1. Many of us get a flu shot each year to protect us from infection with certain viruses. It is quite possible that you could get the flu, regardless of having the injection. Virologists estimate which viruses will come around each season. This is not just a guess on their part, because they study worldwide trends before drawing any conclusions. There is always the likelihood that you could become infected with an influenza virus that was not included in that year's vaccine. Also there is the possibility that you don't develop antibodies properly against those viruses. This explains why some individuals contract the common childhood diseases more than once. Lastly, there is the slim chance that the virus used to make the vaccine was not totally inactivated. So, the yearly flu shot is certainly not a guarantee that you will avoid contracting an influenza virus during that season, but it is one of the best defenses we have at this time.
2. Using monoclonal antibodies for immunization is passive immunity. Passive immunity is sometimes referred to as temporarily induced immunity. The body does not develop memory monoclonal antibodies since these are mouse antibodies. Thus, the monoclonal antibodies are recognized as foreign and removed from circulation. Therefore, they are not kept around long. Since the body recognizes IgG from one's own immune system as *self*, these are not destroyed and the response to these antibodies is long term.
3. Before tissue transplants are performed, the patient (recipient) and potential donor must be tested to evaluate their histocompatibility. In humans, the major histocompatibility complex (MHC) is the human leukocyte antigen (HLA). Such compatibility is necessary to prevent or mitigate rejection of the tissue by the recipient's immune system. The cells that participate in the cell-mediated immune response are responsible for rejection of transplanted tissue.

Chapter 37 Possible Responses to *Data Analysis Exercise* Questions

In this study, researchers attempted to make a connection between the presence of various HPV viral infections and the incidence of cervical cancer.

1. At 110 months, approximately 1% of the women who had no cancer-causing HPV infections developed cervical cancer. In this same time interval, 17% of the women who tested positive for HPV16 had cervical cancer.
2. Those individuals that test positive for both HPV16 and HPV18 would probably be best represented by the line showing the incidence of HPV16-positive participants.
3. This study does not show the overall risk of cervical cancer in all types of HPV-positive individuals, because the one line represented all other cancer-causing HPV types combined. The incidence of cervical cancer would have to be investigated individually for each of the other HPV types as well.

CHAPTER 42: ANIMAL DEVELOPMENT

Chapter 42 Lesson Opener

The NOVA video *Life's Greatest Miracle* is a good video to start the discussion of development. It also revisits the concept of fertilization (Chapter 41), master genes (Chapter 10), and meiosis (Chapter 12). http://www.pbs.org/wgbh/nova/body/life-greatest-miracle.html

Chapter 42 Suggestions for Presenting the Material

- There are numerous excellent animated figures available to help you present this material. The animated figures provide excellent visual context for your students.
- Table 42.2 presents a good overview of human development, and Figure 42.17 presents a good depiction of human embryo development. The specifics of each development in each time interval can lead to a "cataloging" approach, which can be alleviated by using the videotape referred to in the Enrichment section.
- The early development of sea urchin embryos is not as difficult to demonstrate as that of the chick. Biological supply houses sell demonstration kits. Timing is a critical factor for viewing all the stages, so you should plan to videotape the sequence.
- When discussing the placenta, be sure to mention that it not only supplies nutrients and oxygen to the offspring, but it also forms an immune barrier. If humans did not have a placenta, the mother's immune system would eventually detect something "foreign" in her body and the fetus would be rejected. The presence of a placenta prevents this from occurring. Notice that animals like the kangaroo have no placenta. Therefore, the offspring is expelled from the mom at an immature stage and must complete development in the pouch.
- Set up a Jeopardy-type quiz for your students, which may help them distinguish between similar terms, such as blastula, gastrula, etc. A template that contains the setup can be found at http://jeopardylabs.com/.

Chapter 42 Classroom and Laboratory Enrichment

- If at all possible, show a videotape or film depicting development of some animal. Because of the dynamic nature of this process and the rapid changes, static photographs are woefully inadequate.
- The topic of prenatal development will be greatly enhanced by the use of a videotape such as *The Miracle of Life*, distributed by Crown Video through retail outlets.
- Discussion of the tadpole's tail change into a tailless frog gives students a familiar visual image of how apoptosis in development works.
- If your budget allows, purchase some prepared slides of the various stages of embryonic development. Slides of a whitefish blastula are often available for this purpose.

Chapter 42 Classroom Discussion Ideas

- Why do you think the number of multiple births has increased so dramatically in the past 20 years? Part of the reason involves reproductive technology, but also women who have children at older ages are more prone to multiple births. Present the multiple-birth statistics from earlier generations with that of the current multiple-birth trends.
- What are the risks of mothers in their 40s and even 50s giving birth? Evaluate both the risk to the mother and the fetus/child. There are even examples of women giving birth to their own grandchildren via the use of their daughters' donor eggs!
- List ways in which the increased number of multiple births results in increased costs to society.
- The incidence of Down syndrome is said to increase with maternal age, especially for mothers over age 40. Based on the information in the present chapter, can you explain why?
- The placenta supplements, or completely replaces, the activity of three organ systems in the fetus. What are they?
- Many communities, and even states, restrict the teaching of human reproduction. Why do you think this body system is singled out over, say, digestion or respiration for such a prohibition?

- When an insect is in the pupal stage, there is seemingly no activity. Some people have even called it the "resting stage"—erroneously! Biochemically and histologically, what is happening during the pupal stage?
- Explain this quote by Lewis Wolpert (1986): "It is not birth, marriage, or death, but gastrulation which is truly the most important time of your life."
- Make a chart detailing the three germ layers and the tissues they become. Include a description of how this differentiation happens.
- Emphasize the importance of the gray crescent tissue.
- Discuss how the study of embryology has changed scientists' views on classification.

Chapter 42 Possible Responses to *Critical Thinking* Questions

1. There is good reason rubella would cause so many of its effects on the fetus during the first trimester. It is during this time that the organs are developing; limbs form, and toes and fingers are sculpted. Growth of the head surpasses that of other body regions. By 8 weeks, all major organs and systems have been formed, which is why the first 3 months of development (first trimester) is when the infection of the rubella virus has the most effect.
2. Oxytocin has a twofold effect in the mother's body. One of its functions is to produce the uterine contractions that result in the baby's birth. The other function relates to the milk let-down reflex while the baby is nursing. If alcohol reduces oxytocin production, the baby will not get as much milk during the nursing process as it normally would.
3. Germ cells have an important property that somatic (body) cells do not. This characteristic is that germ cells have the potential to differentiate into many different kinds of cells in various tissues. This ability becomes evident in the many different types of structures that can be found in a teratoma. Most teratomas are benign (noncancerous), but they can contain many unusual collections of tissue. It is thought that most teratomas are present from birth, although they may be discovered much later in life.

Chapter 42 Possible Responses to *Data Analysis Activities* Questions

Evidence has indicated that there may be a possible correlation between multiple births and the incidence of birth defects. This study in Florida, 1996–2000, set out to gather data to test that hypothesis.

1. The most common type of birth defect in the single-birth group was heart defects.
2. Heart defects were more common in multiple births than among single births.
3. Multiples have more than twice the risk of developing central nervous system defects as compared to single births.
4. According to this study, a multiple pregnancy does not increase the relative risk of chromosomal defects in offspring.

Unit 6 Homework Extension Questions

1. Name three biological effects of climate change?
2. What are the purposes of inflammation and fever?
3. What happens when the immune system does not function properly?
4. What prevents microorganisms from entering the body?
5. What happens after an antigen is detected inside the body?
6. How do vaccines work?
7. Why is timing and coordination of events necessary for normal development of organisms?
8. What regulates dormancy and germination in plants?
9. How does an adult vertebrate develop from a single-celled zygote?
10. Use an example to explain communication through cell-to-cell contact.

11. How are insulin and glycogen used as chemical signals to illicit cellular response?
12. Discuss the role of cytokines to regulating cellular division.
13. What are the benefits and costs of organisms communicating within their species?
14. List three internal signals that coordinate physiological activity?
15. What role does circadian rhythms play in reference to migration?
16. How can the environment change an organism's behavior?
17. What are the effects of participating in mutualism?
18. What happens when species compete for resources?
19. List two similarities in the structure of plants and animals.
20. Explain why humans are composed of more than one organ system.
21. Why is it important that cells, organs, and organ systems are able to interact?

Unit 6 Lesson Closer

Once this lesson is complete, students should have a basic understanding of how organisms maintain homeostasis and respond to their environment. To close the lesson, watch "Positive and Negative Feedback Loops" on YouTube. http://www.youtube.com/watch?v=CLv3SkF Eag. Divide students into groups of three or four. Instruct students that they are to make a 5-minute video that depicts a positive or negative feedback system they have observed. Show completed videos to the class.

Unit 6 Suggestions for Presenting the Material

- The use of figures, graphs, tables, charts, and scenarios are very useful in describing and illustrating the concepts outlined in AP® courses.
- Real-world examples are very beneficial in teaching students the relevance of this material.
- Discussing glands and hormone secretions can become very monotonous. Break up the material by having activities concerning the hormone, function, and gland the hormones are secreted from.
- Students can observe plant behavior if they take pictures of the same plant everyday for a prolonged period of time.
- Emphasize an essential question that you are targeting as you teach the smaller components of the question. Remind students that the intricate details you are teaching will come together to support a big idea in the end.
- Read the article "Evolutionary Biology of Plant Defenses Against Herbivory and Their Predictive Implications for Endocrine Disruptor Susceptibility in Vertebrates" by Katherine E. Wynne-Edwards. http://ehp03.niehs.nih.gov/article/fetchArticle.action?articleURI=info:doi/10.1289/ehp.01109443

Unit 6 Common Student Misconceptions

- Plants are not capable of changing to their surroundings and are unable to defend themselves.
- Plants are not complicated and grow in a very random order.
- Hormones are only in animals.
- The only functions of hormones are to make people moody or irritable.
- DNA has nothing to do with behavior.
- Desertification and deforestation are not that big of a deal.
- Biodiversity is not important.
- A small change in an ecosystem will not affect the functioning of the system as a whole.
- The ozone is so far away that human actions will never have an impact on its structure.
- Your immune system is made of only antibodies that "attack" bad germs.

- All small cells are bacteria.
- Flowers are parts of plants that enhance their physical appearance with no apparent function.
- A plant consists of a flower, stem, root, and leaves.

Unit 6 Classroom Discussion and Activities

- Homeostasis and Mechanisms of Weight Regulation
  - In this activity, students will investigate how negative feedback mechanisms function to maintain homeostatic balance using a recently discovered system involved in body weight regulation as a model.
  - http://www.utsouthwestern.edu/media/other-activities/251265weightreg.pdf
- Homeostasis and Exercise
  - Students will monitor vital signs to understand mechanisms for maintaining homeostasis during exercise.
  - http://bhhs.bhusd.org/ourpages/auto/2010/4/5/45437230/Homeostasis%20and%20Energy%20Lab%20%202009-2010.pdf

Unit 6 Concept Reinforcement Labs

- Take students outside to a forested area to observe growth patterns in plants. Students should observe spaces between plants as well as direction of growth.
- Students can research different immune system deficiencies and diseases when discussing homeostasis and structure of the immune system.
- Have students conduct a 15-minute observation of a living organism outside the classroom window. Have students list current living conditions, behavior that is allowing that organism to adapt, and physical adaptations.
- Students can research a specific gland and tell not only the function and secretion, but also potential defects that can result.
- Have students photo document a plant growing under different environmental conditions. Then ask the student to infer the differences in the pictures based on its environment.
- Watch "Gravitropism" on YouTube. Have students record and discuss their observations. http://www.youtube.com/watch?v=8SkKuwbmR5Y&feature=related
- Allow students to perform a plant dissection where they are required to remove and label each part of a plant, including each part of the flower.

Unit 6 Case Studies

- Chemical Eric—The Clicker Version
  - A case about the complexity of hormonal control
  - http://sciencecases.lib.buffalo.edu/cs/collection/detail.asp?case_id=503&id=503
- What Happened to 28 Days?
  - A clicker case about the human menstrual cycle
  - http://sciencecases.lib.buffalo.edu/cs/collection/detail.asp?case_id=602&id=602
- Hot and Bothered
  - A case of endocrine disease
  - http://sciencecases.lib.buffalo.edu/cs/collection/detail.asp?case_id=606&id=606
- A Case Study Involving Influenza and the Influenza Vaccine
  - A case about immunity and defense
  - http://sciencecases.lib.buffalo.edu/cs/collection/detail.asp?case_id=326&id=326

- Abracadabra
    - Magic Johnson and anti-HIV treatments
    - http://sciencecases.lib.buffalo.edu/cs/collection/detail.asp?case_id=270&id=270

Unit 6 AP® Practice Essays

1. Compare prokaryotic and eukaryotic cells. Discuss the evolutionary relationship between these two groups of cells.
2. You stub your toe. It swells and is painful. The swelling brings white blood cells to the injured area and the pain makes it difficult for you to move around. Is this an example of positive or negative feedback? Justify your answer.
3. Explain phototropism and how it helps plants to adapt and survive to the environment.

Lesson Outline: Unit 7: Ecology

Correlates with the 15th edition book, Chapters 43–46

- **AP® Biology Big Idea 2:** Biological systems utilize free energy and molecular building blocks to grow, reproduce, and maintain dynamic homeostasis.
- **AP® Biology Big Idea 4:** Biological systems interact, and these systems and their interactions possess complex properties.

Brief chapter summaries

Chapter 43 ("Animal Behavior") is in Unit 6 of the textbook, and functionally acts as the bridge between talking about how organisms function within themselves to how organisms interact with the environment to make an ecosystem (Unit 7).

Chapter 44 ("Population Ecology") explains various ways that populations are studied and some practical applications for that study.

Chapter 45 ("Community Ecology") describes how populations interact with each other in a habitat, and the consequences of shifting community dynamics.

Chapter 46 ("Ecosystems") is one of the culminating chapters in the textbook, bringing together energy exchange between trophic levels in an ecosystem and matter cycling between and within the biotic and abiotic parts of ecosystems in the biogeochemical cycles.

CHAPTER 43: ANIMAL BEHAVIOR

Chapter 43 Objectives

- Information can be communicated between organisms.
- Exchange of information among organisms sometimes elicits a reaction.

Enduring Understanding 3.E: Transmission of information results in changes within and between biological systems.

Essential Knowledge 3.E.1: Individuals can act on information and communicate it to others.

Chapter 43 Warm Up Questions

1. List one example (not human) of an animal communicating with another animal.
2. Explain why a plant may grow a tough, dense shell around eggs laid within the plant tissue.
3. Do you think plants communicate? Why or why not?

Chapter 43 Lesson Opener

We have seen that animals respond to different stimuli. Often animals respond to stimuli that we cannot even detect. For example, it is said that a shark can smell a few drops of blood in an area the size of a swimming pool. For each of our special senses (vision, smell, taste, and hearing), name an animal whose ability to detect stimuli may be more acute than ours. How is this ability adaptive?

AP® is a trademark registered by the College Board, which is not affiliated with, and does not endorse, this product.

Chapter 43 Suggestions for Presenting the Material

- It might be interesting to start out your discussion of animal behaviors by talking about particular animal actions and see if the students can make a guess as to how they are derived. Start out with some easy ones; for example, a monkey learns which lever to press to get a reward. As your discussion continues, get into some more confusing behaviors, such as why a dog circles around before it sits. This will enable the students to see that it is not always so easy to determine the cause or method by which a behavior originates. This may spark their interests in learning more about animal behaviors.
- If possible, show some slides illustrating animal behavior. Many of our students are primarily visual learners and they may understand the actions better by actually viewing them.
- Be sure to stress the specific definitions of classical and operant conditioning so that the students are clear on the differences.
- There may be students in the class that have taken their dog to obedience class. Have them describe the process and what type of learning it utilized.

Chapter 43 Classroom and Laboratory Enrichment

- For those students in the class with some animal experiences, have them name behavior that seems based on instinct. Does there seem to be any advantage to this behavior at one time? Is there any advantage to this behavior now? If not, why does the behavior persist?
- When training animals to learn a behavior, what method works best? If you provide positive reinforcement, how do you determine the most effective reward?
- Humans record habit memory in an area of the brain known as the *corpus striatum*. Is this area of the brain present in other animals that exhibit the habituation type of learned behavior?
- Can you think of other specific courtship displays in other animals? In some species of lizards, for example, the female releases hormones, which prepare her for reproduction, only after she is stimulated by the male's visual display.
- Has anyone viewed the television program *Meerkat Manor*? Have the meerkats on the program displayed any of the behaviors discussed in the chapter?
- The text shows that groups of animals, such as lions, are not more effective hunters in a group. Why do you suppose they often continue this behavior in spite of its lack of success?
- Place an ant farm in the classroom. They are inexpensive and fairly easy to maintain. Observe the functions of the different groups of ants in their eusocial community.

Classroom Discussion Ideas

- Discuss the role of pheromones in insect management. Traps can be devised containing female pheromones and a sticky surface. When the males enter, expecting to find females of their species, they are trapped. Some farmers also use pheromones to simply encourage damaging insects away from their crops by placing pheromones in a distant location. Why are pheromones not used more often by commercial exterminators or regular homeowners?
- Discuss how the interaction between the nervous system and environmental cues might shape and produce behavior.
- What role do genes play in behavior?
- What are the different ways that animals communicate? How might these be adapted to different environments? What are the consequences for noise pollution on the behavior of animals?
- How do male and female mating strategies differ?
- How is cooperative behavior important to social animals?

- From the studies performed on birds, it seems as though they have a window of opportunity to learn to sing a species-specific song. Do humans have this same time period during which they must learn language? This may be hard to prove or disprove since there are few examples of humans developing without early exposure to language. Do you think this could affect an individual's ability to learn a second language more easily early in life?
- Can you think of other examples of a fixed-action behavior of instinctive behavior like that of the cuckoo and the foster parent? Do humans exhibit any behaviors that seem similar?
- Does your household pet exhibit any of the behaviors noted in this chapter? Which of their behaviors appear instinctive? Which ones are learned behaviors? Do they initiate any communication signals to you or other animals? Can you easily interpret what some of these signals mean?
- From the bright coloration of many male birds' feathers, we can assume that birds can discern colors and that they are used as a visual display for prospective mates. Research some additional specific courting behaviors in different species of birds.
- Recent research is demonstrating that personality is common in species other than humans. What are some of the personality traits students have observed in their pets? How might one go about studying personality in animals?
- The chapter mentions the fact that chimpanzees use sticks to extract termites from their mound. Can you think of other similar primate behaviors utilizing primitive tools?
- Discuss the different levels of prenatal care given to various animal species.
- Often the alpha male of a wolf pack is the largest individual. Can you see a genetic advantage to passing those genes along? Do humans keep some of these same characteristics in mind when they select a mate?
- Investigate the study involving the natives in Papua New Guinea and infanticide. Can you find other cultures with a similar practice? Can you see any increase in this type of behavior in ghetto areas where the daily stress is high?
- Oxytocin is sometimes called the "cuddle hormone." Do you think this nickname is accurate, considering that it may involve the monogamous practices of voles?
- How do insects make sounds by rubbing their legs together? What is the purpose of these sounds?

Chapter 43 Possible Responses to *Critical Thinking* Questions

1. The act of moths flying toward the light seems to be an instinctual behavior. These types of actions can remain in a species even though they no longer serve a purpose, or in this case are actually disadvantageous. I would equate this behavior to that of a dog who spins around in several circles before lying down. Both of these behaviors are based on instinct and persist long beyond their usefulness.
2. Since the nonbreeding Damaraland mole rats are not related to those that breed, the purpose of their behavior cannot be genetic in nature. The researchers feel that the actions of the nonbreeding mole rats are ecological, meaning that it enables the mole rats to survive more successfully long term in their current environment.

Chapter 43 Possible Responses to *Data Analysis Exercise* Questions

Many species have ways to protect themselves from predators including mimicking other species, as is the case with the peacock butterfly. Upon opening its wings, two large eyespots are revealed. This experiment seeks to determine whether the eyespots startle a potential predator or serve to mimic the eyes of a predator of the bird hunting the butterfly.

1. When the eyespots were visible, approximately 14 birds gave an alarm call.
2. When the eyespots were hidden, only one or two birds sounded an alarm.
3. These data support the hypothesis that the eyespots serve to mimic the eyes of a predator that hunts the butterfly's predator.

Chapter 43 Homework Extension Questions

1. What are the benefits and costs of living in social groups?
2. Name some potential problems with a modification to a flowering plant that renders it unable to produce a flower.
3. Why would a predator be at an advantage if it could mimic signals given by its prey?

CHAPTER 44: POPULATION ECOLOGY

Chapter 44 Objectives

- Population trends change with the level of variation in a population.
- Examine the different characteristics that are used to describe a population.
- Discuss the factors that determine the size of a population and its growth rate.
- Determine the environmental limits on population growth.
- Discuss how ecologists study life history patterns of different species.
- Discuss how predation affects life history traits.
- Examine the factors responsible for the rapid increase in human population size.
- Determine the factors that will affect future changes in the human population.

Enduring Understanding 4C: Naturally occurring diversity among and between components within biological systems affects interactions with the environment.

Essential Knowledge 4.C.3: The level of variation in a population affects population dynamics

Chapter 44 Warm Up Questions

- What regions of the earth have the highest human population growth?
- What factors contribute to these high rates?
- What are some of the negative consequences of continued human population growth and the continued decline of other species?

Chapter 44 Lesson Opener

Divide the area surrounding your campus into quadrats for population studies. Examine the distribution of species on each plot of land. Determine the population density of an organism in your quadrat. Can you identify if the population distribution is clumped, uniform, or random? Identify factors that you believe might influence species distribution (some examples might be nutrient availability, amount of sunlight, moisture, and manmade disturbances). What changes could you initiate to influence the speciation in your quadrat?

Chapter 44 Suggestions for Presenting the Material

- After presenting a population concept, mention the studies from the text that serve as environmental evidence. Have the students predict the outcome of the study. This will require that they fully understand the meaning of the concept and will encourage participation. If the class does not agree on a single outcome, take a vote before revealing the experimental results.
- Ask students if they can think of another environmental study that relates to the population definitions studied in this chapter.
- Name some various organisms and ask the class to determine whether they best represent a type I, type II, or type III survivorship curve.
- The website http://www.peterrussell.com/Odds/WorldClock.php provides a current census of the world's population on a running counter. You can even illustrate how much the United States and world populations have increased during your lecture! Use the population parameters at the start of class and at the end of class to have students calculate the current human population growth per hour.
- Students will perhaps be most interested in studies of population as they affect human population growth. Examination of the changes in worldwide patterns of population growth since the turn of the century will highlight the overwhelming need for humans to find ways to control the global population growth rate. The inclusion of human population growth studies in this chapter allows students to see that we are not exempt from the rules and limitations that govern all populations.
- Exploration of population ecology also sets the stage for interesting discussions of the socioeconomic impacts of population growth and the ethical questions related to regulating population growth.

Chapter 44 Classroom and Laboratory Enrichment

- Have the students represent the individuals in a population. Designate part of the room as a river, a mountain, etc. Initiate artificial changes in the environment, such as a drought, a flood, introduction of new predators, new food sources, etc. How would this impact both the distribution and the size of your imaginary population?
- Graph the rates of population growth for several of the nations of the world. Discuss reasons for the differences between nations.
- Show examples of age-structure diagrams for human populations of different nations. What conclusions can be drawn about these nations' economies by examining the diagrams? Are the students able to predict which age-structure diagrams came from which countries?
- Show a copy of a life table as published in an entomology textbook. These are especially good examples because they show differential death rates in the various stages of metamorphosis.
- Using the equations for expressing population dynamics, work through examples of problems taken from an ecology textbook.
- Have several students do a plot sampling in a forest or field. It may be difficult to count organisms, but they could easily tally the plants in that specific area. In addition, they could determine if the distribution of plant is random, clumped, or near-uniform.

Chapter 44 Classroom Discussion Ideas

Research the events that occurred on Easter Island and discuss the following points:

- Research a more in-depth look at the events that occurred on Easter Island.
- What were the factors that determined the extinction of the human population on Easter Island? Were these factors density-dependent or density-independent?

- What changes could have been initiated to prevent the population destruction at Easter Island?
- Can you name any other populations that are similar to that of Easter Island?
- Examine the situation on Roanoke Island. How does this differ from the events on Easter Island?
- How do scientists estimate the number of human beings the earth can support? What factors make these estimates unreliable? What are the implications of uncertainty in these estimates?
- Pick a species of animal that is prevalent in your area. Have the class discuss the best and most accurate method for determining its population size. Would a capture–recapture method be practical for this population?
- The class members could discuss the advantages and disadvantages of both r-selection and K-selection reproduction methods. Make a list of organisms that fall into each category.
- Some organisms (bamboo and cicadas) reproduce only a single, brief interlude during their life span but produce a large number of seeds or offspring during this short period of time. Some other plants have seeds that have the ability to remain dormant prior to germination. How do these patterns benefit a species?
- How has modern medical care changed the survivorship curve for humans since the turn of the century? How might the AIDS epidemic alter these curves particularly in some African nations?
- What factors should determine the carrying capacity of the planet for humans? Do these actually operate? Do others operate instead?
- List several resources humans depend on in our daily lives. How many of these resources are unlimited? For those that are not, what should be done to ensure that they are not depleted?
- Name several well-known species whose population numbers are in a decline. Are the factors that determine this population density-dependent or density-independent? Are they an example of an r-selected or a K-selected reproductive method? Have manmade disturbances influenced this population? What could man do to reverse this trend?
- Obtain some data on specific growth rates of bacteria from a microbiology class. Calculate the doubling time for that specific species.
- What are the ethical complications with limiting family size? Do you think nations with high rates of population growth should set a legal limit on the number of children a family may have? Do you feel that birth control can be mandated?
- How do you expect the population of China to be impacted by the government's incentives to control family size? What problems can you foresee for the future since they now produce such an inordinate number of male offspring?
- What environmental factors might influence various human phenotypes (skin color, size, rate of maturity, etc.)?
- How might the current trend toward delayed childbearing change the population growth in the United States? Why would an increase in parental age change the population structure?
- What density-independent factors influence the size of insect populations? Could we use this knowledge to devise nonchemical methods of insect control?
- What is the relationship between the size of the offspring and its number per reproductive event?
- Some mammal populations often breed in the springtime and larger males seem to be the desired mates. What are the advantages of these practices to the population? Do humans show any of these same biases?
- Why will the human growth rate of a country like Mexico continue to soar for many years to come even if stringent birth control measures are started immediately?

- Can you envision any solutions to the economic crisis that will result when the baby boomers reach old age?
- Which type of population distribution would you likely find in plants? How does this differ in animals?
- Think of some examples of introduced species in an area (e.g., the kudzu plant, zebra mussels, rabbits, etc.). Did the introduction of these species upset the native populations? How did man's intervention cause the problem? Could human actions help solve the problem of the invasive species?
- Examine U.S. census data for your county or state at http://www.census.gov. See if you can calculate the per capita growth rate from these data. If pertinent figures are not available, make up some sample examples for the calculations.
- As an example of exponential growth, examine the following facts concerning reproduction in rabbits:
    - Rabbits sexually mature at 5 months.
    - Their gestational period is 31 days.
    - The average litter size it six rabbits.
    - A female can have eight litters per year.
    - For this study, we are assuming the following: start with one female, there are no deaths, and the population is 50% female and 50% male.
    - The end result after 1 year is 1850 rabbits!
    - Graph the figures to see if these data demonstrate exponential growth.
    - Can you see why rabbits became an invasive organism when they were introduced in Australia?

Chapter 44 Possible Responses to *Critical Thinking* Questions

1. The selection pressure for survival from a predator like the cichlids may have influenced a duller appearance with the male guppies. Flashier males might attract more unwanted attention from the predators. When released from the predation pressure, the males may have been selected for "gaudiness" because the females prefer that appearance. Sexual selection could now be operating in the male guppy populations after they have been released from predation pressure.
2. The age-structure diagram on the left appears to be from a pre-industrialized nation with a rapid growth rate. Individuals are more likely to have many children if they are needed to help with agriculture. Therefore, the largest groups of individuals are the younger ones. The death rate is quite high from birth on, which may be the result of poverty and malnutrition. It also looks as though this population does not have many medical advances since only a very few live until old age. With this type of rapid growth rate, density-dependent limitations, such as food and water, may be a great factor in the future. The age-structure diagram on the right resembles a developed country, with a lower birth rate, perhaps due to better access to reproductive health and birth control, as well as greater educational and career opportunities for women. In addition, this diagram suggests the availability of modern medicine, since many individuals live until old age.

Chapter 44 Possible Responses to *Data Analysis Activities* Questions

Researchers decided to study the populations of marine iguanas in the Galápagos Islands. Select iguanas were marked and their health, survival, and habits were studied over time.

1. There were greater numbers of marked marine iguanas on Santa Fe Island at the time of the first census.
2. There was no change in the population size on Genovesa Island between the first and second census. The population on Santa Fe Island dropped dramatically between the first and second census, from ~180 to ~70.

3. If an adverse event had affected both islands, the populations of marine iguanas would have plummeted on both islands. Since the population on Santa Fe Island was the only one that decreased dramatically, it appears as though the oil spill caused the decline.

CHAPTER 45: COMMUNITY ECOLOGY

Chapter 45 Objectives

- Environmental changes in an environment result in changes among growth and homeostasis in biological systems.
- Competition among organisms changes population size and distribution of species.
- Ecosystem distribution changes over time.
- Biodiversity ultimately affects interactions within the environment.
- Stability of an ecosystem may be determined by the diversity of species within.
- Populations interact with each other in communities.

Enduring Understanding 4.A: Interactions within biological systems lead to complex properties.

Essential Knowledge 4.A.5: Communities are composed of populations of organisms that interact in complex ways.

Chapter 45 Warm Up Questions

1. What is a community?
2. What happens when species compete for resources?
3. How do predator and prey populations change over time?

Chapter 45 Lesson Opener

- Select any environment and take a look at the predator–prey relationships within it. Draw a Venn diagram to see what food sources may overlap between two predators. How does this overlapping affect those particular prey populations?
- Examine a vegetated area on campus or in an area nearby. How are the plants in the area competing for resources? Suggest some ways in which competition has shaped the plant community.

Chapter 45 Suggestions for Presenting the Material

- Explain the philosophy behind each of the types of species interactions. For example, how does a parasite derive the maximum amount of benefit from the host without causing its death? How can it maintain its lifestyle without being detected by the host?
- The elaborate species interactions described in the chapter offer an opportunity to discuss coevolution. Examples such as the array of yucca species in Colorado, each pollinated exclusively by one kind of yucca moth species, emphasize the point that individuals don't evolve, populations do. Students will be able to see many good examples of adaptive traits in this chapter.
- The coevolution of predator and prey is another good example of the impact of one species on the evolution of another. After reading and discussing this chapter, students should understand that communities are shaped by a complex web of many different factors.
- Show some slides of animals in nature exhibiting camouflage, mimicry, and warning coloration. Can the students always spot the organism in the photo?

- Discuss the introduction of Japanese beetles into the United States. How did they enter the country? What control methods have been initiated? Some beetle traps utilize pheromones, but some researchers worry that many crops are destroyed by insects traveling to these traps. How can the use of pheromones be used effectively? Is this a safer method of insect control than the use of pesticides?'
- Not all attempts to use other organisms as agents of biological control have been successful. Discuss some of the pitfalls associated with biological controls (e.g., attack on nontarget species). What are some potential strategies for avoiding such failures?

Chapter 45 Classroom and Laboratory Enrichment

- Look into the chemical relationship between the nodules on legumes and their ability to alter nitrogen. When was this discovered? What particular chemical reactions are involved? Can you envision any way to artificially induce these changes in atmospheric nitrogen?
- Look at some preserved samples of human parasites. How do they enter the host's body? What can be done to prevent their introduction into the body?
- Choose a community, list its species, and categorize them as producers, consumers, decomposers, or detritivores.
- Examine construction sites, flooded river banks, plowed fields, and other places that have been recently disturbed. Can you find several plant species that you would describe as pioneer species?
- Look at any natural setting. Describe the habitat and niche for every organism observed.
- Have the students name other examples of camouflage in nature. After they name a few, show some examples, such as the phasmids (walking sticks).
- Describe patterns of succession at edges of stream beds, rivers, or coastlines.
- Look in your own backyard or in areas that surround your school for examples of resource partitioning.
- Using the observation that grass quickly establishes itself in cracks in the pavement of a highway on which traffic has been blocked for some time, describe what is happening using terms from the chapter.
- Discuss the biological control system that is used to combat infestation by gypsy moths.

Chapter 45 Classroom Discussion Ideas

- Have the class voice their opinions on the ideas of instilling parasitoids in the environment to eliminate fire ants. What else should be considered when introducing a species, such as the phorid flies?
- Do the phorid flies have a deleterious effect on native ant populations or any other naturally occurring insects?
- Why do insects introduced into the United States become such pests when they were not so in their native country?
- Do the phorid flies serve as a vector for any diseases in other species such as man or livestock?
- Do the students have any alternative suggestions for controlling the fire ant population? How would one go about creating a pesticide that would only affect the fire ant population?
- Is there any possibility that insect pheromones could be used to remove the harmful fire ants from the environment? Why are biological controls not used more commonly by the average homeowner?
- Have there been any previous examples of foreign insect infestation that could be studied? Were there any methods that proved effective against the invading insects?
- Are certain plants "born" to be weeds? Or do they achieve that status by human condemnation?

- Investigate some of the adaptations that parasites have undergone to survive in the caustic environment of the host's digestive tract.
- What is the difference between true parasitism and social parasitism? Can the students give examples of both types of lifestyles?
- Why is resource partitioning essential for groups of functionally similar species living together?
- In some areas of the country, wildfires are a natural phenomenon. Should they be allowed to proceed without human intervention?
- Relating to environmental issues, voters tend to vote with their wallets rather than with scientific concerns. How can we stop this dangerous trend?
- What characteristics distinguish a pioneer species? Are pioneer species good competitors against later successional species? Why are pioneer species dependent on the frequent advent of open, disturbed places?
- Discuss predator–prey interactions. Why are the cycles of predator and prey abundance, shown in Figure 45.10, described as idealized? What do you think a predator would do if deprived of its primary prey item? Examine the actual diets of several predatory species; how do these diets change from one month to the next throughout the year? How can environmental disturbances such as fires, floods, climate fluctuations, and insect outbreaks influence the predator–prey cycle? What are some of the other variables that may be overlooked in predator–prey interactions?
- Would you expect competition between two finches of different species to be less intense or more intense than competition between two finches of the same species? Explain your answer.
- The monarch butterfly is orange and black and tastes bad (birds eating them spit them out immediately); viceroy butterflies are almost indistinguishable from monarchs but taste good. Which of these is the model and which is the mimic?
- Why should so-called "good ideas" like kudzu get more scrutiny before being released into a habitat?
- According to your text, "In primary succession changes begin when pioneer species colonize a barren habitat." Are there any uninhabited places left on Earth for pioneer plants and animals to colonize? What would create such a setting?
- There are different types of mutualism. Investigate each of these types and the following examples:
  - Trophic mutualism (derive benefit from resources)—zooanthellae algae and coral polyps
  - Dispersive mutualism (help in distribution of pollen)—plant and insect pollinator
  - Defensive pollinator (protective advantage)—acacia tree and ants
- Review the three types of responses seen in predator–prey interactions. Can the students name examples for each of these models?
- Discuss the methods that plants utilize to deter herbivory. How many examples can the students name?

Chapter 45 Possible Responses to *Critical Thinking* Questions

1. The principal of competitive exclusion best explains the situation when cows are instilled with an increase of naturally occurring bacteria to discourage the growth of an undesirable strain. Since both species would inhabit the same niche, it is reasonable that the well-established species would not allow the other to effectively colonize. This seems much more desirable than feeding the cows antibiotic-laced food for several reasons. First, the antibiotics may destroy the natural flora of the cow's digestive tract along with the harmful strain. This may make the digestive process in the cows more difficult. Secondly, the consumers of the milk or meat from the cow may ingest small amounts of antibiotics. Therefore, we may be creating an atmosphere for antibiotic-resistant organisms to develop in the consumer.

2. If flightless birds have existed for many generations in an island environment, it is not surprising that none develop the ability to fly. There would be no adult birds of that species to encourage that skill. Also, the current structure of the flightless birds is not designed for flight.

Chapter 45 Possible Responses to *Data Analysis Activities* Questions

Fire ants are an introduced species that causes discomfort in many southern states. This study was designed to test the effectiveness of ant-decapitating phorid flies and the microsporidian *Thelohania solenopsae.*

1. The population size of the control fire ants increased over the first 4 months of the study.
2. The fire ants in the two types of treated plots decreased drastically to almost zero over the same time period.
3. If the study ended after 1 year, one would conclude that the biological controls had a major effect on the fire ant population.
4. At the end of the study at 28 months, one could still conclude that the efforts to eradicate the fire ants were somewhat successful. It appears, however, that some of the ants had become resistant to the pesticide by the study's end.

Chapter 45 Homework Extension Questions

1. What factors contributed to the increase in the human population size?
2. What factors affect the types and abundances of species in a community?
3. What determines the size of a population and its growth rate?

CHAPTER 46: ECOSYSTEMS

Chapter 46 Objectives

- Biotic and abiotic factors affect cells, organisms, populations, communities, and ecosystems.
- Each organizational level of a biological system exchanges matter and energy.
- Changes occur within and outside of ecosystems as a result of information exchange.
- Energy and matter are transferred at each organizational level within a biome.

Enduring Understanding 2.D: Growth and dynamic homeostasis of a biological system are influenced by changes in the system's environment.

Essential Knowledge 2.D.1: All biological systems from cells and organisms to populations, communities, and ecosystems are affected by complex biotic and abiotic interactions involving exchange of matter and free energy

Enduring Understanding 4.A: Interactions within biological systems lead to complex properties.

Essential Knowledge 4.A.6: Interactions among living systems and with their environment result in the movement of matter and energy.

Enduring Understanding 4.B: Competition and cooperation are important aspects of biological systems.

Essential Knowledge 4.B.3: Interactions between and within populations influence patterns of species distribution and abundance.

Essential Knowledge 4.B.4: Distribution of local and global ecosystems changes over time.

Enduring Understanding 4.C: Naturally occurring diversity among and between components within biological systems affects interactions with the environment.

Essential Knowledge 4.C.4: The diversity of species within an ecosystem may influence the stability of the ecosystem.

Chapter 46 Warm Up Questions

1. Explain what happens to a freshwater fish if it is placed in a saltwater environment?
2. What is the difference between a biotic and an abiotic factor?
3. Name three abiotic factors that influence an organism living in a forest.
4. What is energy?
5. What is a biogeochemical cycle?
6. How does sunlight affect climate?
7. Define *competition*.
8. Why is biodiversity important?
9. Name at least three things that can cause a change in the number of species in a population.
10. How do human activities endanger existing species?
11. Why is it difficult to establish clear boundaries between ecosystems?
12. List three reasons why ecosystems change over time.
13. What is genetic diversity?
14. Why are small groups of organisms more likely to experience extinction than large groups?
15. How does natural selection maintain diversity?
16. Define *biodiversity*.
17. Why is biodiversity important for the health of an ecosystem?
18. Explain the effects of extinction of one species within an ecosystem.

Chapter 46 Lesson Opener

Speedy: (10–15 minutes)

1. Read *The Blind Persons and the Watershed* available at http://www.fs.usda.gov/Internet/FSE_DOCUMENTS/stelprdb5073114.pdf to the students.
2. Discuss the following questions:
   a. What were the blind persons in the story?
   b. What were their perspectives?
   c. Why do they have those perspectives?
   d. How were you "blind" before reading this story? (What were some things you didn't know prior to reading?)

Extensive: (40–50 minutes)

1. Read students the following prompt:
   Coastal wetlands are an important factor to insure the success of bird migration. Ponds, lakes, and marshes provide food and shelter for traveling birds. Without the wetlands, birds would not have the energy to make the trek from areas as far south as Panama in the case of the Belted Kingfisher. At the time of the European settlement of the United States, there were 215 million acres of wetlands. Today, there are less than 100 million. Besides providing habitats for waterfowl, wetlands help relieve flooding, filter pollutants, and are an integral part of the biosphere. Through increased education of their importance and beauty, children will become responsible stewards of the remaining 100 million acres of wetlands.
2. Draw a large-sized hopscotch course. The course can be drawn on the pavement with chalk or drawn on the sand/dirt with a stick. The squares should be approximately 3' × 3'. The hopscotch course should contain 10 squares.

3. Have the students line up at the beginning of the course. Tell the students that they are birds starting their journey northward. Tell the students that each of the squares represents a wetland between Florida and Maine (It will be more dramatic using a migration path, which includes your state. Specific migration patterns and bird species can be obtained from a bird field guide.). Students are then challenged to migrate northward on the course. They do not have to step on every square; however, they must not go outside the course.
4. All students should be successful in the first migration. Now, tell the students you are a developer. You will destroy two wetland areas in order to build condos. Put an "X" on two of the squares. Tell students to make the migration once again. The students may not set foot on the destroyed wetlands. If they do, they die and thus may not participate in any further migrations. After all students have run through, destroy two more and repeat the procedure. Repeat this until all students fail to make the migration. Try to "X" off the squares in such a way that not all are destroyed but are so far apart students cannot make the jump. This will help with the debriefing.
5. At the end of the activity, ask students the following questions:
    1. Explain why some birds died earlier than others?
    2. Why did the rest of the birds die?
    3. Explain how this game represents migration.
    4. Why did the birds die even though some wetlands remained at the end of the game?
    5. Why is it important to save wetlands in all states?
    6. How do migrating birds depend on wetlands during migration?

Source: http://ofcn.org/cyber.serv/academy/ace/sci/cecsci/cecsci045.html

Chapter 46 Suggestions for Presenting the Material

- Approach the study of the various cycles of an ecosystem by first explaining the necessity of that particular component in the environment. Then examine the deleterious effects of too little or too much of that substance.
- Begin by showing a simple food chain and an example of a representative animal at each trophic phase. Then stress that life and nature are rarely that basic. Show the students an example of a complex food web and discuss the unlikelihood that an animal would have only one food source. Create a disruption in the food web by imagining the destruction of a population of one species of animal. Let the students see that a seemingly small change in a food web could have a dramatic impact on many other species. Students will become aware of the unity that joins all organisms.
- Have the students view the award-winning documentary *An Inconvenient Truth*. This powerful film may help to launch a discussion of the consequences of global warming and the impact that this movie has created worldwide.
- Use as many local examples of ecosystems as possible in discussions, demonstrations, and lab work. While it may seem overly simplistic and sometimes inaccurate to identify and describe the different levels of an ecosystem, students should see that it is useful because it helps us to understand the functioning of an ecosystem as a whole. Such ecosystem descriptions provide a valuable baseline against which we can measure the effects of changes.
- Stress the fact that, in various sections of the chapter, the authors show how humans have altered the natural ecosystems and their functioning.
- Use overhead transparencies to present the biogeochemical cycles.
- Invite a representative of the EPA, the state coastal commission, or another environmental group to address the class.
- Have students build their own food web using the online resource available at http://www.gould.edu.au/foodwebs/kids_web.htm.

- Go around the room and have the students name examples of producers, consumers, detritivores, and decomposers that would be in the area surrounding your campus.
- See if your students can devise a food chain (or better yet, a food web) for organisms in your area. This will serve to show them how complicated and intertwined these relationships are.

Chapter 46 Classroom and Laboratory Enrichment

- **AP® Investigation # 10: Energy Dynamics**
- **AP® Investigation #12: Fruit Fly Behavior**
- Have some of the students choose a cycle discussed in this chapter and make a poster depicting the stages. Each participant should explain his poster to the class and how human intervention may alter the natural cycle.
- Set up aquatic ecosystems in the lab and monitor them throughout the semester. Identify the trophic levels of the ecosystem and analyze the cycling of materials and nutrients within it. In what ways are the aquatic ecosystems in the lab similar to, or different from, a real aquatic ecosystem?
- Discuss the primary productivities of different regions of the United States. How can human intervention change primary productivity?
- Each student could undertake a study near their home of an environmental issue regarding an upset of one of the natural biogeochemical cycles. Are there any reasonable solutions to this issue? If so, write to your senator or congressman expressing your views.
- Test various water supplies for the presence of nitrogen and phosphorus. Are they within the recommended standards? If not, how far do they deviate from a normal level? Use a nursing or anatomy text to determine if an excess of either substance would cause any short-term or long-lasting effects?
- Determine the pH of water samples from several local bodies of water in your area. How much does the pH vary? If you have suitable equipment at your school, you can use a Global Positioning System to create maps of exactly where you obtained your samples. Since you are somewhat familiar with the area, can you find a reason to explain these differences in pH?
- Fill an aquarium with water and organisms from a nearby pond or lake. Add amounts of phosphorus and nitrogen to mimic eutrophication. What effects do you see in your artificial ecosystem?
- Analyze local soils to determine the mineral and organic contents.
- Divide the class into small groups. Have each section debate the pros and cons of different energy sources. Each group should be prepared to discuss the merits and drawbacks of one of the following energy sources: oil, coal, natural gas, hydropower, solar power, nuclear power, or other alternatives. Can the class come to any consensus about the safest method or combination of methods?
- Assuming that a flat 10% of the energy in one trophic level is conserved to the next, and that the producer represents 100%, calculate what percent is received by the marsh hawk in Figure 46.4.
- Write down the models of cars driven by some of your classmates. Research online to see if you can determine the emissions from each of these models. When you announce the results to the class, will that influence anyone's choice for their next vehicle?
- Visit a sewage treatment plant. Discuss the biological steps involved in the treatment of your local sewage. In what way could sewage treatment be improved?
- Research the steps that will be undertaken to convert the astronauts' urine into drinkable water in space.
- Collect some soil from what appears to be a detrital food web in a forest or field system. Can you identify any organisms as decomposers and detritivores? Do you imagine that some of the organisms performing these duties may be too small to see?

- Contact the epidemiologist at your local hospital. Inquire how they handle their dangerous biomedical wastes. In the past, it was sufficient for a hospital to simply contract with a company to dispose of hazardous wastes. Now hospitals are required to know the pathway of the dangerous trash after it leaves the building. This makes it clear that all parties involved are responsible for proper environmental procedures.
- Design a simple survey containing a few concise questions relating to environmental concerns addressed by this chapter. Have each student poll 10 fellow students outside of the classroom. How do your peers feel about environmental issues? Do they consider this an urgent matter to address? And most importantly, would they be willing to put their money where their mouth is?
- Look online for an example of a biomass pyramid for an ocean ecosystem. How does it differ from those seen in Figure 46.8?
- Have your students bring in containers from detergents and fertilizers that they use in their own homes. Read the labels to see if there is a significant difference between them. Would your students be willing to switch to a brand that contains less phosphorus, even if it costs more?

Chapter 46 Classroom Discussion Ideas

- Does anyone see the ironic dichotomy that Morgan City in Louisiana celebrates a shrimp and petroleum festival on the same day? The production of oil may be emitting pollutants into the air that are increasing the greenhouse effect. The result of the increasing greenhouse gases, as we know from the text, is contributing to global warming. As global warming intensifies, polar ice caps will melt and, therefore, raise the water level. The lowlands of Louisiana, with some of the lowest levels of coastal area in the United States, may disappear at an alarming rate. If this continues at this current rate, the shrimp festival may have to be moved to a city farther inland, where it will have become the new coastline!
- In addition, scientists have discovered that an increase in ocean temperature does not cause more hurricanes, but the ones we do have will worsen in intensity. As we know from Hurricane Katrina, the devastation from a strong hurricane can result in the displacement of millions of people, and many billions of funds are needed to attempt to repair the area.
- The Louisiana bayou area may be one of the areas that will show the early effects of global warming. We need to observe the damage that is being done now and address it dramatically before it is too late. The nations of the world must band together to consider global warming as a very serious worldwide problem that must be aggressively attacked with manpower and financial means.
- One consequence of global warming that may not be immediately evident is the spread of tropical diseases. As the Earth warms, the boundaries of where tropical diseases can exist may expand. Diseases such as malaria and dengue fever may eventually become common in the United States.
- Discuss the rapid deforestation of the rain forests. How does this impact the biogeochemical cycles that we have addressed?
- Discuss what would happen to an ecosystem if all of the producers disappeared. What would happen to the ecosystem if all of the consumers, decomposers, or detritivores disappeared? Can you think of examples of ecosystems in which any of these events has occurred? How can an ecosystem recover from such a disturbance?
- Show examples of various biomass pyramids. Can the students match them with the proper ecosystems? How would a biomass pyramid from the open ocean differ from the one in Figure 46.8?
- Describe several trophic levels of a typical ecosystem and ask the students to arrange them in the correct order.
- How and where do humans fit into a food web?

- The pesticide DDT is just one example of a substance that undergoes biological magnification as it travels through an ecosystem. Can you think of others? (One possible example is the movement of strontium 90, a byproduct of nuclear testing in the 1950s, through the food web.)
- Look into articles about the effects of biological magnification in Lake Erie? Which animals are affected? What steps have been undertaken to remedy some of the environmental issues there?
- What do you think were the first producers to evolve on Earth? Do we know what they might have looked like?
- It is expected that sea level rise will be the most immediate and serious consequence of climate change. What are the two primary drivers of sea level rise? Are there communities already being impacted? What are the long-term social consequences if sea levels continue to rise and meet future projections?
- Examine the composition of several plant and garden fertilizers. What are the major ingredients? What components provide the "miraculous" growth rates?
- Why is the term *food chain* rarely used when describing actual ecosystems?
- Why are humans at the top of nearly every food web of which they are a part? Are they ever at any other level? How has man's role in food webs evolved through the centuries?
- What would the personal and ecological advantages be to humans if they were to eat "lower down" on the energy pyramid?
- Why does a pyramid of energy usually narrow as it goes up?
- Research some alternative fuel sources for automobiles. Why do you feel these are not readily available?
- Is it environmentally wise to rely on large quantities of nitrogen-rich fertilizers for crop production? What are some alternatives? Discuss the pros and cons of commercial fertilizers.
- Try to locate examples of energy pyramids from various locations. It may be impressive for students to see the loss of energy at each phase. Contrast the amount of energy lost by endotherms and ectotherms, and explain why this difference occurs.
- Why is a pyramid of biomass a more accurate representation of an ecosystem than is a pyramid of numbers?
- Investigate the source of your home water supply. If applicable, see if you can find the name of the aquifer that supplies the water.
- Why does normal rain water have a pH of around 5.6? (*Note:* A pH of 7 is neutral.)
- How have coal-burning power plants changed in the past 20 years?
- Perform some research on the ozone layer above the Earth. Where did they first discover a hole in the ozone layer? Can anything be done to repair the damage that has occurred?
- Tropical forests are highly productive ecosystems, incorporating extremely large amounts of carbon and other nutrients into plant material. Yet, when cleared of vegetation, such areas make poor farmlands. Why?
- About 500,000 trees are needed for Sunday newspapers for Americans. Is this necessary? When will we no longer be able to afford such extravagance? How would the environmental and economical impact change if you received your paper online?
- In each of the biogeochemical cycles, indicate the route each component takes in recycling. What invisible component is not recyclable?
- Make a list of renewable and nonrenewable energy sources. Which method or combination of methods would you recommend?

- Draw a timeline indicating the dates of the different legislative actions relating to car emission standards.
- Ask students to make a list of ways in which they could modify their own lifestyles and behaviors to reduce environmental pollution.
- Make sure your students understand what happens to the energy that is lost at each trophic level. Who loses more energy, ectotherms or endotherms?
- There are lots of terms in this chapter that must be defined. Make sure your students know these before they attempt to learn the overall concepts.
- Contact the local weather service to determine which tests are performed on collected rainwater.

Chapter 46 Possible Responses to *Critical Thinking* Questions

1. The vegetable gardens in Maine and Florida would differ greatly based on climate (e.g., rainfall, temperature, latitude, length of growing season, soil content, and perhaps proximity to the ocean), depending on their exact locations. All of these factors would greatly influence what types of crops would be best suited to each region. Florida would have a longer growing season, so I would expect the garden in that area to have a higher annual productivity.
2. The website listed takes you to a site that is sponsored by the U.S. Geological Survey department. At this website, you can click on a map to increasingly zoom down to about the county level. It will then give you information about your particular watershed, such as water quality, location of groundwater inventory sites, etc. In addition, it offers detailed definitions of all technical terms for the layperson's interpretation.
3. One would not be able to test the air in bubbles in frozen ice for phosphorus, because phosphorus does not occur naturally in that form. The best way to check phosphorus levels from the past would be to test the levels in sedimentary rock, since this is the major reservoir for phosphorus.
4. If you increased the number of nitrogen-fixing bacteria in an aquatic ecosystem, it could result in an algal bloom and eutrophication. The additional nitrogen would encourage an overgrowth of algae, which may cause a hypoxic atmosphere for the animals that reside there. This eutrophication would lead to an increase in carbon production and an increase in carbon accumulation.
5. If the mycorrhizal fungi were used to provide plants with phosphorus and nitrogen, the risk of eutrophication would decrease dramatically. The runoff from the fields would not contain the excessive amounts of nutrients that would be seen in an area exposed to large amounts of fertilizers. Some research should be performed before such a widespread plan is initiated to ensure that the fungi in large amounts would not cause any significant disruption in the overall balance of nature.

Chapter 46 Possible Responses to *Data Analysis Activities* Questions

Scientists are attempting to show that human activities have impacted the amount of carbon dioxide in the atmosphere. In Antarctica, there is thick ice that can be used to measure the carbon dioxide level over time.

1. The highest carbon dioxide level between 400,000 and 0 A.D. was 300 ppm.
2. During this same time period, the carbon dioxide level never reached the level attained in 1980.
3. The trend in the carbon dioxide level for the 800 years prior to the advent of the industrial revolution was fairly constant, with some slight variations. After 1800, the carbon dioxide level began to rise dramatically.
4. The rise in the carbon dioxide level between 1980 and 2013 was much higher than the rise between 1800 and 1975.

Chapter 46 Homework Extension Questions

1. Give an example of how a biotic and an abiotic factor can affect an ecosystem.
2. How can a cell be affected by an abiotic factor?
3. Explain how predator–prey relationships can change the homeostasis of an ecosystem.
4. What is the trophic structure of an ecosystem?
5. How does energy flow affect food chains and food webs?
6. How does energy flow through ecosystems?
7. What factors will affect future changes in the human population?
8. How does community structure change in response to resource partitioning?
9. How do predation and herbivory influence community structure?
10. How do ocean currents arise and how do they affect regional climates?
11. What are some ways that pollutants directly harm living organisms?
12. What can individuals do to reduce their harmful impact on biodiversity?
13. Does evolution occur in recognizable patterns?
14. Name three things that can affect population dynamics in an ecosystem.
15. How does a population's genetic diversity become reduced?
16. Explain what might happen to an ecosystem if mutations and natural selection no longer existed.
17. Explain the importance of a keystone species in an ecosystem.
18. How can a desirable trait in a population ultimately decrease biodiversity?

Unit 7 Lesson Closer

Once this lesson is complete, propose a human disturbance (such as an oil spill), and instruct students to list possible effects that will occur not only on oceanic ecosystems, but also on a global level. Follow up by allowing students to research a specific human disturbance and present their findings to the class.

Common Student Misconceptions

- A small change in an ecosystem will not affect the functioning of the system as a whole.
- Abiotic factors have no influence on biotic factors and vice versa.
- There is little communication between organisms, especially if they are not of the same species.
- Predator–prey relationships ensure a balanced population.
- Once an ecosystem is defined, it always stays the same size.
- Biodiversity has no effect on the functioning of an ecosystem.
- Humans have no effect on ecosystems.
- Once an ecosystem has been damaged, there is no way to repair it.
- Trophic levels are not related to animal communication.

Classroom Discussion and Activities

Additional Labs

- Animal Behavior
  - In this lab, you will be studying animal behavior by working with terrestrial isopods commonly known as pillbugs.
  - http://www.charleszaremba.com/aplabs/animal.html

- Population Genetics and Evolution
  - Using the class as a sample population, the allele frequency of a gene controlling the ability to taste the chemical PTC (phenylthiocarbamide) could be estimated.
  - http://www.charleszaremba.com/aplabs/popgen.html
- Dissolved Oxygen and Aquatic Primary Productivity
  - In this exercise, you will measure and analyze the dissolved oxygen (DO) concentration in water samples at varying temperatures. In part B, you will you will measure and analyze the primary productivity of natural waters or laboratory cultures using screens to simulate the attenuation (decrease) of light with increased depth.
  - http://www.charleszaremba.com/aplabs/dissolved.html
- Competition
  - In this lab, you will use a computer model based on the Lotka–Volterra competition equations to gain a more intimate understanding of the factors that can influence the outcome of competition in a simple environment. You will also use a computer model based on Tilman's resource competition model to evaluate how resource dynamics can influence the outcome of competition.
  - http://www.cnr.uidaho.edu/wlf448/comp1.htm

Concept Reinforcement Labs

- Have students recreate natural disasters on a small scale and allow them to brainstorm and try ways to reverse the effects.
- Identify different types of pollutants and have each student write a one-page research paper on one type. After the papers are written, require each student to present their findings to the class.
- Direct students to research endangered species in their area. Instruct each student to write a proposal that will facilitate conservation of those species.
- Contact your local sanitation district to obtain materials to sample water in the school's watershed and contribute to analysis of water quality through submission of your findings.
- Visit the site http://www.concord.org/sites/default/files/projects/er/materials/TeacherGuide_Activity4-TXMO-final.pdf and lead students through the activity entitled "Changes in the Environment."

Case Studies

- The Wolf, the Moose, and the Fir Tree
  - A case study of trophic interactions
  - http://sciencecases.lib.buffalo.edu/cs/collection/detail.asp?case_id=453&id=453
- Threats to Biodiversity
  - A case study of Hawaiian birds (populations)
  - http://sciencecases.lib.buffalo.edu/cs/collection/detail.asp?case_id=449&id=449
- Who Set the Moose Loose?
  - Trophic interactions in the Greater Yellowstone Ecosystem
  - http://sciencecases.lib.buffalo.edu/cs/collection/detail.asp?case_id=502&id=502
- What Is a Species?
  - Speciation and the maggot fly
  - http://sciencecases.lib.buffalo.edu/cs/collection/detail.asp?case_id=551&id=551
- Speak Up!
  - Mini cases in language
  - http://sciencecases.lib.buffalo.edu/cs/collection/detail.asp?case_id=497&id=497

AP® Practice Essays

1. Analyze the organization of an ecosystem by describing trophic levels. Include how trophic levels are divided, how energy flows through them, and the relationships between different levels.
2. Discuss four types of direct interactions that involve two species. Include an example of each relationship.
3. The following food web represents the organisms of a large community. The arrows show the direction of energy flow in the system. For example, the arrow points from the harp seal to the cod because the seal obtains energy from the cod. Categorize the animals according to their trophic level and means of obtaining energy.

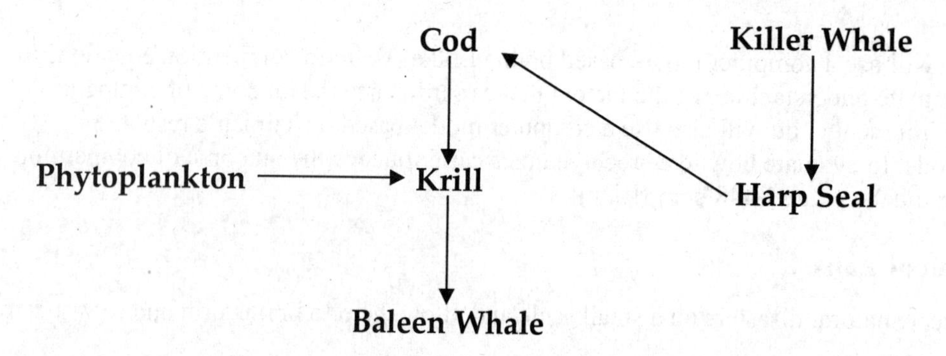

Common Problems for Teachers and Students

Common Problems for Teachers

The biggest challenge facing teachers is the newly designed AP® Biology curriculum. This guide is written to be a helpful tool to the new AP® curriculum and help teachers navigate what is within the new framework. This guide reflects the AP® Biology curriculum at the time it went to press. It may have changed; please see the website advances in AP® for the latest information.

Pacing

Pacing is a perpetual problem for AP® Biology teachers. We are challenged with teaching a great deal of information within a limited time. Most textbooks that are suitable for this level contain 50 chapters or more of very detailed information that teachers feel obliged to present to students. Staying on track and finding ways to cover all of the chapters are challenges that even the most experienced teachers face. One of the goals of the revised AP® Biology curriculum is to decrease the amount of information and give teachers time to go deeper into the material.

The new AP® Biology curriculum is divided into big ideas, enduring understandings, and essential knowledge. Teachers will now have choices about what examples they would like to use in order to illustrate concepts. Under essential knowledge, it will state a number of examples that teachers can use to illustrate an idea or it will state that students must demonstrate understanding of each of the examples. Therefore, not all of a textbook is required to be taught, but teachers must make sure the required examples are taught. This manual will help guide teachers on what material is within the new curriculum framework and also specifically note where possible examples are at the teacher's discretion.

How do you face this new challenge?

Plan your course for the school year during the summer. Sit down with your school calendar and a blank calendar. Cross out all of the days you know you will not have class—holidays, assemblies, school-sponsored events, etc. Count how many days there are remaining from the first day of school until the day of the AP® Biology exam. Use this number to determine how many class days you should spend on each of the four big ideas.
Here's an example:

> At the school where Mary teaches, there is a total of 180 days for the school year. Because the AP® Exam is on the second Monday in May, 15 of those 180 days are after the exam, so she has 165 days of school before the exam. Throughout the year, there are 2 days for special testing (PSAT, etc.), 5 days for assemblies, and 3 days for teacher professional development. This is a total of 10 days that Mary's class will not meet, giving her a total of 155 days for teaching, laboratory, and testing AP® Biology. Using the method explained above, Mary should allocate those teaching days as follows, according to the course outline as given by The College Board. Many schools have even fewer actual teaching days, so it is important to count and plan accordingly in the summer.
>
> Big Idea #1 The process of evolution drives the diversity and unity of life = 17% or 26 days:
>
> - Evolution is the change in allele frequencies within a population over time.
> - Common ancestry between organisms.
> - Changing environments cause evolution to constantly occur in nature.
> - Origin of life.

AP® is a trademark registered by the College Board, which is not affiliated with, and does not endorse, this product.

Big Idea #2 Biological systems uses energy and molecular building blocks to grow, reproduce, and maintain homeostasis = 30% or about 46 days:

- All life requires energy.
- All living organisms must maintain homeostasis in a changing environment to survive.
- Feedback mechanisms (positive and negative) are used by organisms in order to maintain homeostasis.
- Changes in the environment cause changes in homeostasis.
- Timing and coordination of biological processes are essential for growth and reproduction.

Big Idea #3 Living systems retrieve, transmit, and respond to information essential to life processes = 30% or about 46 days:

- Genes provide a basis for all living things.
- Cellular and molecular mechanisms are used to regulate genes.
- One source of genetic information is from the processing of genetic information.
- Cell communication involves signal, transduction pathway, and cellular response.
- Communication results in changes between biological systems.

Big Idea #4 Biological systems interact, and these interactions possess complex properties = 23% or about 35 days.

- Interactions between organisms and the environment cause complex properties.
- Working together and against each other causes changes in biological systems.
- Diversity among and between organism's affects environmental interactions.

When planning lessons for these days, Mary will also need to include time for testing, time for laboratory, and time for review just before the exam.

As you plan your course, remember to allow some flexibility for unexpected schedule changes, adjustments in the school calendar, as well as inclement weather days. Once school starts and you begin teaching your course, stick to your plan as much as possible; however, you will find that you will make adjustments according to how quickly the students grasp the material, for the inclusion of additional laboratory exercises or other activities, and your own strengths and interests. Realize that even under the best of circumstances, you may not be able to cover all of the topics in the detail you would like, though avoid "curriculum creep" that ends up short-changing the final chapters. Plan ahead for topics you will cut if need be. Focus on making your class the best biology course it can be and those things that you do teach…teach them well.

Adequate Time for Completing the Required Laboratories

Finding adequate time to complete the required laboratories in 45–55 minutes can also present a challenge for AP® Biology teachers. While some of the laboratories are more time-intensive than others, they each require attention to detail for the understanding of the underlying concepts, designing a control experiment, and the analysis of the results. The new labs use an inquiry-based approach to investigate meaningful questions about the real world. The labs are integrated into the curriculum and align with the scientific practices. In implementing inquiry-based investigations, teachers can update their existing labs or use the manual from the college board. Students should have ample practice being able to experimentally investigate self-generated questions and use the results to draw conclusions. A minimum of 8 out of the 12 labs are required (two per big idea). As in the past, 25% of class time should be devoted to laboratory time, and teachers are encouraged to complete more than eight labs if possible. Remember that you will want your students to practice writing scientifically, as well, which takes up class and/or homework time.

Here are some suggestions for completing the labs:

Incorporate the simpler lab activities into the lecture. For instance, the Blast Lab from Big Idea #1 can be done in class and discussed as a class over multiple days or done at home by students and discussed in class.

Prepare students to come to class ready to complete the lab. As teachers, we often want students to obtain a total experience by doing all parts of the lab. However, student preparation can take up valuable class time that would be better spent doing the lab. There are numerous AP® Biology lab-based instructional videos on YouTube that students can watch the night before to help them design the experiment outside of class. That way they will be familiar with the lab design and can come to class ready to carry out their designed experiment. Teachers can focus on helping make sure students have designed a controlled experiment and that the students understand what their results mean.

Schedule separate or additional time for lab. If possible in your school schedule, try to schedule a separate course for AP® Biology Lab. This will allow designated time for both lecture and laboratory that will allow adequate time for each. If it is not possible to schedule lab as a separate course, require that students occasionally come in outside of the regular school day to complete the labs. This will be a challenge at first, but as time passes and students understand the expectations of the course and appreciate the value of extra "contact time," it will become less of a problem, especially when supported by the administration of your school.

Multi-Day Labs

Many of the inquiry labs can be performed over multiple days in class. Students may try an experimental design and find that the results do not make sense. They can come back the next day and use the knowledge gained from the previous day's experience to adjust the experimental design. Many of the inquiry labs can take 2–4 days of class time to complete.

Obtaining Supplies and Equipment for the Required Laboratories

There are several companies that provide kits that can be purchased for each of the 12 labs, as well as refills for kits already purchased. These kits are certainly helpful but can be expensive if you have budget constraints.

Obtain alternate supplies. Many of the materials needed for the labs can be purchased inexpensively from your local grocery or hardware store.

Distribute a "wish list" to parents. Parents are often willing to donate supplies that cannot be accommodated in your budget. At the beginning of the year, give parents a list of what you need for your lab. You may want to include things such as sugar, hydrogen peroxide, string, French fry cutters, yeast packets, peas, cotton, graph paper, gloves, etc. Many parents have careers that are science related and may be able to arrange for the donation of supplies that are being replaced or upgraded in their labs or offices, such as test tubes, beakers, sphygmomanometers, gel-electrophoresis equipment, spectrophotometers, or microscopes. Obtaining donations will allow you to focus your budget on those materials that can only be obtained from scientific supply companies.

Develop partnerships with local colleges or universities. Depending on your school's budget, it may take several years to obtain all of the materials and equipment that you need to complete all of the labs. Lab 6: Molecular Biology is of particular concern to many AP® Biology teachers because the gel-electrophoresis equipment can be expensive. This need not be a deterrent to having students complete the lab. Contact the biology departments at your local college and/or university. Inform the department chair of your need and request their assistance in developing a partnership with one or more faculty members who would be willing to assist you by allowing your students to come to their facility to complete the lab (depending on the size of your class) or by allowing you to borrow their equipment. The science faculty at your local colleges and universities may also be willing to donate used equipment to your classroom as they upgrade the equipment in their labs.

Adequate Time for Testing

Finding enough time to test students in ways that prepare them for the AP® Exam can be tough, particularly for teachers who have shorter class periods or who teach on a block schedule. Teachers strive to prepare students for the AP® Exam by simulating it in their classrooms on test days. In doing so, many teachers are tempted to administer tests over a 2-day period, giving a substantial multiple-choice section on day 1 and free-response questions on day 2. While the intent of this practice is good, it can be time-intensive and result in the loss of valuable teaching days over the course of the school year.

Avoid the temptation to "over-test" students. Students can be equally prepared for the AP® Exam by taking 1-day tests during the year. In a 55-minute class period, it is possible to simulate AP® Exam testing conditions by asking 35 multiple-choice questions (45 seconds per question; 26 minutes total), 2 grid-in questions (2 minutes per question; 4 minutes total), and 1 multi-part free-response question (22.5 minutes). If you have shorter class periods, you may choose to give the equivalent of a quarter of an AP® Exam for each test (22.5 minutes of approximately 25 multiple choice questions and 1 grid-in question, and 22.5 minutes for a 10-point free-response question or 3 short free-response questions. Each 22.5 minutes is worth 50% of the final test score). Be strategic in the questions selected for the test. Instead of asking many easier questions on a single concept, choose questions of moderate difficulty that require students to think about the concept on a deeper level or link concepts together, or strive for a balance of easy and difficult questions. Write or select a free-response question on a topic that is not addressed in the multiple-choice section.

Explaining Difficult Concepts

AP® Biology is a college-level course taught in high school. Since lecture is an important part of the college experience, most AP® teachers fall into the habit of daily lecture in order to prepare students for college. Lecture is indeed important. There are additional important ways to get students to understand concepts.

Use animations. Cengage Learning has included short videos and animations as part of their ancillary packages as well as their websites. Take advantage of these resources and use them in your classroom and/or assign as homework. Encourage your students to incorporate the publishers' resources into their daily study routine. There is a wealth of animations that can be found on the Internet as well. Animations bring to life those concepts that are difficult to visualize in a two-dimensional picture.

Use some flipped lessons. For some of the lectures, have students watch a video lecture that you make yourself or you find online. When they come to class, you can get them right to work applying what they learned to a case study, lab, or some other activity designed to engage them in deeper thinking. While they work, you can wander around the room asking students or groups of students questions to informally assess their understanding and answer their questions.

Have students act out cellular processes. Getting students physically involved in learning complex processes helps them to more effectively understand and remember them. Here is an example of a skit that can be used to help students understand transpiration in plants.

> The doorframe of the classroom represents a stoma. The door itself represents the guard cell. The open door represents the open guard cell and thus the open stoma. The converse is true for the closed door. Close the door of the classroom. Explain to the students that it represents a closed stoma. Ask 8–10 students to volunteer to be water molecules. Have them line up in front of the door. Remind the students that water is cohesive. Have the water molecules hold hands to demonstrate the cohesiveness of water. Open the door. The stoma is now open. Gently guide the first water molecule out of the stomata. Because the water is cohesive, the rest of the water molecules will be pulled toward the stomata. Continue until the students understand the concept. To further emphasize the point, have an additional two to three students join the line from further back in the room, to demonstrate water traveling up the xylem from the roots.

A variation on that theme is to break your class up into two or more groups and challenge them to design their own skit to model a particular process. The group with the clearest, more accurate, and most easily understood skit "wins." Similar skits can be used for other processes, such as glycolysis, the electron transport chain, double fertilization, translation, and more. There are no limits to what you and your students can do to help their understanding.

Use hands-on activities. Small, hands-on activities are also helpful to students. Often, inexpensive toys can be used to illustrate concepts students have difficulty visualizing. For example, snap beads can be used to demonstrate how dehydration synthesis reactions form peptide bonds between amino acids. Dominoes can be used to demonstrate the propagation of action potentials in neurons. Giggle toys can be used to illustrate how energy is passed from ATP to other molecules. Capillary tubes can be used to demonstrate the role of cohesion and adhesion on the movement of water through the xylem of plants. For some processes, you can use a more student-centered approach by giving students a variety of materials or craft supplies and have small groups come up with the best way to model or make an informational classroom poster. Contests work well for motivating students to be as accurate and clear as possible, along with being fun.

What to do when you're absent from school. The absence of the teacher does not have to result in the loss of a day of learning for students. The days the teacher is away from class can be used wisely. If the absence is planned, it can be used as a testing day. Alternately, it can be used as a lab day for those lab activities that may not require as much teacher supervision. Work time for many of the student-centered activities outlined earlier can be done in the absence of the teacher, if needed. By using teacher absence days in this manner, continuity will not be lost, and the class will be ready to proceed when you return.

Reviewing for the AP® Exam

Ensuring that students are adequately prepared for the AP® Biology Exam is one of the most important, yet challenging priorities of many teachers. There is no right way to review with students. Many methods can be used to review the plethora of information that students have learned over the course of the school year. Some in-class review is strongly encouraged when possible.

It is important in reviewing with students to ensure that they are learning to make connections and think conceptually. To accomplish this goal, teachers should help students link big ideas together, such as homeostasis. Students should be able to make comparisons among organ systems and among organisms. Students should also be able to link cellular processes with physiological and/or ecological processes for example:

- macromolecules, enzymes ➔ digestive system (optional example), cellular respiration
- cellular transport ➔ urine formation (optional example), neuronal transmission
- cellular respiration ➔ respiration, carbon cycle
- photosynthesis ➔ transport, carbon cycle
- communication ➔ within cells, between cells (nervous, endocrine systems), and between organisms (pheromones, coloration, behavior)

Of course, evolution is the overarching theme that should be emphasized throughout the school year.

Review sessions can be held in class and outside of class. Some teachers have out-of-class review sessions after school or on weekends. Review sessions can be held at school, in students' or teachers' homes.

Strategies for Review in Class

1. Review Exams: Students can be given practice exams on the information taught in each quarter or semester.
2. Practice Exams: Teachers can administer practice exam (available to teachers who have created an audit account on The College Board website).
3. Fun and Games: Teachers and students can re-enact skits from earlier in the year to review difficult concepts/processes. Additionally, students can create and play games: Jeopardy, Who Wants to Be a Millionaire, Weakest Link, or Vocabulary Bingo.
4. In groups, have one student randomly open the textbook to three different places. The other student(s) then describe(s) how those three concepts connect to one another, giving as much detail as possible.
5. Laboratory Review: Teachers can ensure that students understand the objectives of the labs by reviewing the background, procedure, and analysis for each exercise. Additionally, students should practice writing free-response answers to lab-based questions.
6. Test-Taking Strategies: Many teachers help students with test-taking hints during review time. Topics to cover can include how to effectively use the 10-minute reading period before writing their answers.
7. Remind students that there is no penalty for guessing.
8. A formula list will be provided for the exam. The formula list will be published on the Advances in AP® website for students to use on practice exams and throughout the course.
9. Calculators may be used for both sections of the AP® exam. Starting with the 2018 administration of the exam, students are allowed to use a four-function, scientific, or graphing calculator.

Common Problems for Students

Finding Time for Daily Study and Using Study Time Wisely

Students have many academic and extracurricular commitments that place demands on their time. Even the most motivated students struggle with maintaining a consistent study schedule.

What can teachers do to help? Recognize that students want to succeed. Set reasonable expectations for their class preparation. Depending on the class schedules of your students, expecting them to read ahead may be overwhelming if they are taking multiple AP® courses and involved in extracurricular activities. Encourage your students to go home each day and spend 30–40 minutes solidifying their understanding of the material taught that day. Also, encourage students to review during the weekend by reading over the sections that were discussed in class during the week and go over their notes. They should fill in any information from the textbook that may help clarify information in their notes. Doing so will give the students clear, identifiable milestones to reach each day. They will understand the foundations, enabling teachers to more effectively build on previous knowledge. Teachers may want to provide incentive by giving short quizzes each day at the beginning of the school year until students realize what works best for them and develop a routine.

Understanding Difficult Concepts

For most students enrolled in AP® Biology, your class will be the first time they have been expected to think on a conceptual level and learn information on a college level. Many concepts are simply difficult for students to grasp the first time.

What can teachers do to help? Acknowledge that the material is difficult and be supportive of your students' efforts. While there will be a few students who just "get it," some students will struggle regardless of the amount of study time they devote to learning biology. Provide tutorial sessions 30 minutes before or after school as often as possible. Encourage students to attend, and affirm their improvements when they do. Many students realize the benefits of participating in study groups for the first time in their schooling. As students begin to grow and understand better, they will become motivated to continue attending and will persuade their classmates to attend as well.

Encourage students to use the supplemental materials that are ancillary to your chosen textbook. Most textbooks recommended for AP® Biology have CDs, study guides, and websites available for student use. If students do not have direct access to these materials personally, provide them access in the classroom by having them available for students to use in the classroom or check-out for use at home in the evening. The more often students interact with the material, the better they will understand it.

Learning the Language of Biology and Proper Use of Vocabulary

Understanding the language of biology is crucial to students' understanding of the material. Often, their understanding is hindered by the lack of vocabulary understanding.

What can teachers do to help? Avoid the temptation to require students to know every vocabulary word in the chapter. When planning your lessons for a particular chapter, determine which words are essential for students to know in order to understand the material in that chapter. Helping students to recognize common prefixes and suffixes used in biological words is an important part of increasing their understanding of the concepts and relationships. Provide students with a list of common prefixes and suffixes at the beginning of the year, along with their definitions. Encourage students to keep the list where it is easily accessible so that they can refer to it throughout the year as they learn in class or study at home.

Seeing the Big Picture: How Do the Details Fit Together?

There is a wealth of information that students learn in AP® Biology, and it is frequently difficult for students to think conceptually as they are learning the details. They may be accustomed to learning groups of facts without knowing how the facts fit together or how they fit into the "big picture."

What can teachers do to help? Identify the "big picture" idea or "take-home message" for students at the beginning of class each day. If students know the goal of their learning for that day, it will be easier for them to put the details in context as you teach. Have students practice writing answers to free-response questions throughout the year in preparation for the AP® Exam. As students learn more information, give them questions that require them to link together multiple parts of the curriculum. These questions may be given as in-class assignments, take-home essays, and/or given on tests. Allowing students to practice answering conceptual questions will give them the skills and confidence they need to fully understand the material and to achieve maximum success on the exam in May.

Review Activities for the AP® Biology Exam

Goals

- Encourage the students to study and prepare for the exam.
- Think of ways to build confidence.
- Think of ways to reduce stress and anxiety.
- Remind students that the cognitive skills they learn and practice in preparation for this exam will be transferrable to other courses and life experiences, regardless of the eventual score on the exam.

Activities

1. Use multiple-choice and free-response questions on tests given during the year to condition students for the format of the exam.
2. Discuss free-response writing tips, exam directions, exam format, and the grading of the exam prior to the exam date.
3. Give students the multiple-choice and grid-in portion of the exam included in this guide during a 90-minute-timed period, so students can experience the time span they have to answer. This could be done before or after school if the school is not on a block schedule. Block scheduling would allow time for the practice exam during the regular class period.
4. To practice the multi-part free-response questions, have students write or outline the free-response questions from old exams or the questions included in this guide.
5. Read over grading standards from old exam free-response questions (these will be like the multi-part free-response questions) with the students. Allow students time to practice grading their own free-response answers, their classmate's answers, or sample answers using the standards. This will build their confidence and enable them to realize best organizational strategies for their own answers in the future.
6. Have students develop their own color-coded review packets, which might include such topics as chemiosmosis; the Krebs cycle; the Calvin cycle; DNA replication, transcription, and translation; important molecular structures; nerve impulse transmission; and so on. You could have them swap review sheets and ask them to look for new, missing, or erroneous content in their partner's sheet. Have them discuss with each other afterwards.
7. If new material is being taught up to the date of the exam, begin having review sessions of the previously taught material before or after school beginning about 6 weeks before the exam. Each review session should be about an hour long.
8. If class time runs out before the topics can be covered, divide the students into groups and have the groups prepare presentations.
9. Give weekly review quizzes over specified material before or after school.
10. Have students create flash cards or short quizzes online and review these as sponge activities before class begins each day. Alternatively, put up questions from this guide or the released AP® practice exams relevant to the content of each class to begin each day.
11. Periodically use the first 10 minutes of class to allow the student's time to play concept development or vocabulary games.

AP® is a trademark registered by the College Board, which is not affiliated with, and does not endorse, this product.

12. Have groups of students make concept maps using words relating to each chapter and/or Big Idea. Hang up the concept maps when done to allow students to compare and discuss.
13. During the year, make videos of the labs as the students do them and watch them during time dedicated for lab review. There are also many websites that have virtual labs and are a good review for students.
14. Go over the objectives for each of the labs. Focus on the lab process skills. Be sure the students can design a scientific experiment or take given data and analyze the scientific experiment. Many free-response questions ask students to design labs and analyze data.
15. The week before the exam, concentrate the review on the material that was studied during the first semester (or at the beginning of the semester if this is a one-semester course).
16. Have students reread the chapter summaries and go over the end-of-chapter questions found in the texts.
17. Break students into small groups. Assign each group one Big Idea to write in large letters in the middle of a piece of newsprint or poster board. Hand out bags containing slips of paper with the Enduring Understandings (EUs) and/or Essential Knowledges (EKs) on them. The challenge is to assign the appropriate EUs and/or EKs to the appropriate Big Idea, grouping them by related concept and/or Illustrative Example.
18. Divide the students into groups and assign each group a major topic to review. Each group will prepare a short presentation with handouts and visuals of the assigned topic. Examples of topics for review with the corresponding lab activities include the following:

Topic	Big Ideas
Organic Molecules	
Enzymes	Enzyme Catalysis
Energetics and Metabolism	
Cells	
Membranes	Diffusion and Osmosis
Cell Organelles	
Photosynthesis	Photosynthesis
Cellular Respiration	Cellular Respiration
The Cell Cycle: Interphase, Mitosis, Cytokinesis	Mitosis and Meiosis
Meiosis	
Mendelian Genetics	Mendelian Genetics
Molecular Genetics	Molecular Genetics
Evolution	Population Genetics and Evolution
Plant Diversity	
Plant Anatomy and Physiology	Transpiration
Animal Diversity	
Animal Anatomy and Physiology	Physiology of the Cardiovascular System
Animal Behavior	Animal Behavior
Ecosystems and Biomes	Dissolved Oxygen and Aquatic Primary Productivity

What to Do after the AP® Exam?

These are suggested activities for AP® Biology classes after the exam is completed in May. It is important to keep the students learning, but in a less stressful manner. Do not let them think the exam is the end of their learning experience.

Suggested Activities

1. Current Events "Speed Dating": Assign short or longer articles from secondary science sources such as *Scientific American* or *Science News*. Have students identify and summarize the Background (purpose, etc.), Methods, Results, and Conclusions (including future directions and/or caveats) for homework. In class, have students share what they learned and how it relates to what they learned in class in 2-minute time intervals + 1 minute of questions before being the listener for 3 minutes. Then swap partners and repeat. Use the last 5–10 minutes of class for students to write down as many things they can remember about what they learned.
2. Biomimicry Project: Biomimicry is the engineering principle of looking to nature to help inform how humans can solve a problem. An example of this is Velcro, which was modeled off of the shape of cockleburs stuck to the inventor's dog. Have students use biomimicry to come up with their own invention, build a prototype of it or an aspect of the invention using common household or craft materials, and give a 10-minute PowerPoint pitch to "potential investors".
3. Field Trips: Visit classes and laboratories at nearby research universities, aquariums, science museums, local hospitals, the zoo, or botanical gardens. These trips can be scheduled after school if field trips during the school day are not allowed.
4. CPR Training: Contact the local Red Cross office and have a volunteer come to your school to teach CPR and first aid. These lessons usually take about a week to complete. At the completion of the course, your students will be certified in CPR.
5. Laboratory Enrichment Activities: Have students conduct some of the laboratory investigations you did not have time for during the school year before the exam. Examples might include dissections, forensics labs, genetics labs, and ecology labs. Each lab activity could take several days. Students could also use their independent investigative skills learned and assessed throughout the AP® curriculum to conduct and present their own experiments.
6. Get connected with the local elementary and middle schools. The amount of time invested in these activities can vary. Examples include the following:
   a. Have your students visit the local schools and teach science lessons.
   b. Have your students write and illustrate science books on various topics (e.g., dinosaurs, pond life, insects, spiders, etc.). Laminate and bind the books and give them to the schools. Your students could visit the schools and read the books to the students explaining the science involved.
   c. Assign each of your students to mentor an elementary student when the group goes on a field trip to the zoo, aquarium, science museum, etc.
7. Science Symposium Night: Invite parents and other students to a science symposium. Have students give 15 minutes of information in the form of PowerPoint presentations or poster presentations on particular topics in science. If the students have done science fair projects, this would give opportunities to share what they learned from doing the work for the projects.

AP® is a trademark registered by the College Board, which is not affiliated with, and does not endorse, this product.

8. Technology: The amount of time invested in these activities can vary. Students also gain familiarity with search engines. Examples include the following:
   a. Have students create PowerPoint projects on biology topics that can be used to help your classes next year or to help teach a concept to the introductory biology classes.
   b. Have students do Internet research on particular topics and present this information in class.
   c. Have students use the Internet to do a "Web Essay" on an open-ended topic.
9. Notebooks: Use class time for students to organize their work in AP® Biology into a notebook, including information from the labs. In order to do this correctly, they must look over all the material they have collected for the year. This might take a couple of class days. Collect the notebooks and grade them. Return them to the students to use for reference for their college biology classes. Better yet, this can be done before the exam to help them organize their materials for more effective studying.
10. Scavenger Hunt at the Zoo: Work with a representative from the local zoo to arrange a scavenger hunt there with your students. Have prizes for the winners. This does not have to be a class trip together. Students may go to the zoo on their own time and bring their answers back to you.
11. Movie Day: Now is the time to show those movies you never had time to show. Some excellent suggestions include *GATTACA*, *The Double Helix*, *Lorenzo's Oil*, and *Gorillas in the Mist*. Each of these movies will take two or more days to show and discuss.
12. History of Biology Day: Have your students role-play a famous day in history. Each student will have a specific role in acting out an event in biological history. For example, the day Watson and Crick realized they had discovered the structure of DNA. Your students will have to research the individual they are portraying and actually convey that information as they act out the part.
13. Cancer Mini-Documentary or Public Service Announcement: Have students research tumor development due to sun exposure. Have small groups of students create mini-documentaries or public service announcements on skin cancer. Post these videos on the school's website or broadcast on the school's news program.
14. Commercials for Next Year: Assign each student group a different unit that was studied during the course of the year. Have students create video commercials that will serve as introductions to each unit for next year's AP® Biology class. The commercials can preview the unit, highlight interesting assignments, and give hints for the tough stuff. You can also create podcast previews that next year's students can subscribe to via your course on the school's website. If you do not have the technological resources to complete this activity, try assigning movie posters to advertise each unit (with very specific guidelines on content and presentation) that you can hang in the room next year at the beginning of each unit of study to spark student interest.

Movie and Book Studies

1. *An Inconvenient Truth* and *State of Fear*
   Have students complete a book and movie study on how scientific data are interpreted and used to support arguments. Global warming is a relevant topic in today's media and most students have an opinion or position. Use the movie *An Inconvenient Truth* and the book *State of Fear* by Michael Crichton to investigate scientific bias, research funding, and learn about global warming. Students can watch the film in class and discuss the methods of presenting data and tracking information. Students should read the novel at home during the film discussion. The lesson produces numerous teachable moments and valuable discussion. Students learn to be critical viewers and readers. Assign a position paper in which the students use what they learned while reading and watching about the global warming debate. They should take a position on the issue and use the reading and information from the video, along with additional research, to support their case. Make sure they do not turn in a paper full of facts on global warming. They should take a position and defend it with data.

2. *The Hot Zone* and *Outbreak*
   The book *The Hot Zone*, by Richard Preston, highlights the immergence of several viruses in the human population. The book looks at the course of viral disease, classification of viruses, and host range. This is a good opportunity to discuss viral structure, classification as living or nonliving, and immune response. You can also explore topics of globalization and the spread of disease. The recent public concern over the possible spread of avian flu is a good way to open discussions on handling the rapid spread of microorganisms due to global travel. Studying the epidemiology behind tracking an outbreak and the real-life problem-solving and critical-thinking skills needed for this type of scientific investigation are excellent discussion points. You can show clips or short sections of the movie *Outbreak* that correlate to the day's book discussion.

3. *And the Band Played On*
   This is a fictionalized movie on the epidemiological investigation, obstacles, and disagreements about the appearance of AIDS and later discovery of HIV. This movie provides an opportunity to discuss the true obstacles that can impede or delay scientific discoveries. Have students recall other scientists whose work met with numerous obstacles before being validated (Mendel, Darwin, and Franklin). Debate how the situation may have been handled differently and the difficulties faced by all. You can assign roles and have students re-create a conference that may have been held to discuss different viewpoints. Have students research their roles and role-play based on their research during the conference. This should help to illustrate that point of view and bias can play a role in scientific decisions. The movie is a fictionalized account, which gives the opportunity to ask students to research the real story.

Portfolio Project: Stem Cells

If you have 3 weeks after the AP® exam, students can work on portfolio-style projects on stem cells, genetic engineering, genetically modified foods, etc. This gives students the opportunity for extended research, demonstration of understanding at various cognitive levels, and will spark a curiosity into cutting-edge science that often gets overlooked or glossed over in the curriculum due to time constraints. This is a sample portfolio for stem cells. One good resource is via the California Institute of Regenerative Medicine (CIRM): https://www.cirm.ca.gov/our-progress/stem-cell-education-portal-landing

Stem cells are the focus of breakthrough medical research. These cells are also a source of controversy among scientists, doctors, researchers, politicians, and legal experts. The portfolio will be presented in a binder in the order listed next.

1. Stem Cell Foldable: Create a matchbox-style foldable with tabs for what stems cells are, where they are formed, why they are important, and how they are formed.
2. Concept Map: Create a concept map using the terms you encounter in your research of stem cells. *Stem cell* should be the central term with branches for *controversy*, *uses*, *description*, *research*, and *future*.
3. Newsletter: Create the front page of a newsletter that gives general information about stem cells, their current uses, and future possibilities. Address both sides of the stem cell debate without bias. Alternatively, you may create the front page of a newspaper reporting major breakthroughs in stem cell research. The articles must be original and can include current achievements or inferences on future breakthroughs based on your research.
4. Experimental Design: Use an experimental design worksheet to design an experiment that tests one of the possible uses of stem cells discovered in your investigation. Write a logical, thoughtful hypothesis; decide on controls; and with your variables in mind, decide how you will collect data. You will not be conducting this experiment, simply designing it.

5. Position Paper: The culminating piece of your portfolio is the position paper. After completing all your research and investigating stem cells, you will need to choose a position on the future of stem cell research. In the paper, you will need to defend and support your position using facts you encountered. This paper is not a summary about stem cells. Your opening paragraph should introduce the idea of stems cells and state your position. You may use one paragraph of your paper to give a brief overview of stem cells, but the remainder of the paper should focus on supporting your stated position.
6. Work Cited Page

Group Discussion Activity: Ecological Crisis in Ecuador

This activity is designed to show how seemingly simple environmental issues involve numerous stakeholders and concerns. No easy solutions are found in such situations and compromise may be the only way to satisfy as many perspectives as possible. Inflexibility, although it may satisfy a minority of stakeholders in the short run, often results in greater harm.

This activity closely mirrors a similar situation that occurred in Ecuador between the Cofan peoples of the Ecuadorian rainforest and oil companies. A book was written about the situation titled *Amazon Stranger: A Rainforest Chief Battles Big Oil* by Mike Tidwell. There was also a television documentary filmed called "American Chief in the Amazon" shown on the A&E network (*Investigative Reports*) and produced by Bill Kurtis.

It may be helpful to have a large map of Ecuador available for the class to see. Think ahead about small suggestions you might need to interject if negotiations stall or deviate. This assignment is purposefully open-ended. Encourage groups to think of creative solutions to the problems and find common ground. Make sure it is clear to each group that their "minimums to negotiate" are not to be revealed directly to the other groups. This discussion is divided into five parts: the indigenous peoples, the World Bank, the Green Voice, the ACME Oil Company, and the Ecuadorian government.

- The Situation: [*Every group should read this paragraph*]
  The ACME Oil Company has recently discovered a large deposit of oil in the remote tropical rainforests of Ecuador. ACME has approached the Ecuadorian government with a plan to build a massive drilling operation in the rainforest along with the construction of a pipeline from the drilling area to a city on the coast with port facilities. The Ecuadorian government is very interested in this project as a boost to their slumping economy, but they can't afford to sponsor the project alone. The World Bank is willing to loan money to the Ecuadorian government but is cautious of investing in unstable countries and unstable situations. The area where the drilling is to take place is inhabited by several tribes of indigenous peoples, who do not support the project as it is currently proposed. A large, international environmental group, Green Voice, has become involved on behalf of the indigenous peoples. Green Voice is also deeply concerned with the destruction of rainforest habitat in Ecuador.

- The Indigenous Peoples: [*Only this group reads this paragraph*]
  Your tribes have lived in this part of the tropical rainforest since your Inca ancestors migrated to the area several centuries ago. Your people continue a simple lifestyle of living off the abundance of food and resources found in the rainforest. There has been a long history of bad relations with local land owners and the Ecuadorian government. All of this has made you suspicious of "outsiders." You strongly oppose this project because it will effectively ruin your traditional way of life. Unfortunately, your tribes comprise a small population and a therefore weak voice in the country's government. Additionally, illiteracy is high and governmental healthcare for your tribes is virtually nonexistent. Your tribes badly need better government services, but they are costly and the government appears to be unable (or unwilling) to fund them. Because you are not good

at negotiating your needs with the government, you have been willing to let Green Voice help you dispute your needs. You've made it clear to Green Voice you are unwilling to lose your way of life by losing the rainforest.

<u>The Indigenous Peoples' minimum to negotiate</u>: You are absolutely not willing to move from the area. Your people will support the project only if your way of life in an intact rainforest can somehow continue.

- <u>The World Bank</u>: [*Only this group reads this paragraph*]
  The World Bank is an enormous financial institution interested in funding large-scale development projects around the world and especially in underdeveloped nations like Ecuador. This is a risky business, however, as many of these developing nations have defaulted on their loans and left your investors with nothing to show for their money. You are dealing with the Ecuadorian government directly in this matter of the ACME Oil Company's drilling project proposal. Several financial reports within the World Bank have confirmed that the Ecuadorian government can pay back the loans if the project nets at least 20 billion U.S. dollars per year. *These reports are highly confidential and the threat of international law suits prevents you from disclosing this figure to others*. The situation is shaky and your institution is very reluctant to risk the loans if the situation between the Ecuadorian government and the peoples of the rainforest does not improve.

  <u>The World Bank's minimum to negotiate</u>: You are willing to fund the project only if the drilling venture is profitable enough to ensure the pay back of the loans. Additionally, you require the Ecuadorian government to guarantee a stable situation in the rainforest. Do not reveal your information about the profits needed to make the project work.

- <u>Green Voice</u>: [*Only this group reads this paragraph*]
  Green Voice has a long record of championing environmental causes around the globe. Your group is especially skilled at advocating for the rights of indigenous peoples. You have approached the tribes of the Ecuadorian rainforest, and they have reluctantly agreed to allow you to help them fight the ACME oil drilling project. The indigenous peoples in this forest have been treated poorly by local land owners and the Ecuadorian government for a long time, and it has made them resentful of those outside of their tribes. Your group is additionally concerned about the environmental destruction to the rainforest by this project. Green Voice recently hired a team of scientists to survey the area and the team discovered numerous endangered, threatened, and unique species of biota found in this part of the rainforest. You are currently researching the possibility of an alternative plan that would reduce the size of the project and spare some of the rainforest. It is unclear if the indigenous peoples, the Ecuadorian government, or the ACME Oil Company would agree to a limited drilling project.

  <u>Green Voice's minimum to negotiate</u>: You must not allow the rights of the indigenous peoples to be ignored in this negotiation. You are also absolutely unwilling to let endangered species perish.

- <u>The ACME Oil Company</u>: [*Only this group reads this paragraph*]
  ACME Oil, like all oil companies, is a business and understands the world through the lens of a business eye. Although recent geopolitical events have resulted in record-setting profits for oil companies, competition could not be anymore stiff. Every project, therefore, must count and be profitable. World oil reserves are dwindling, while demand for petroleum products continues to grow. These factors have led your company to take a risk on the Ecuadorian rainforest. ACME has already spent millions on exploration in the Ecuadorian rainforest and it looks like it may pay off. It's a big project worth as much as 40 billion U.S. dollars a year if the proposal is implemented immediately and at full scale. Internal ACME studies have determined the project

must net at least $10 billion a year for it to be worth the trouble to pursue with the Ecuadorian government. The Ecuadorian government is behind you 100%, but they are having trouble with funding and meeting the demands of indigenous peoples located in the forest. If the government can't stabilize the situation, your company will be greatly concerned about the safety of its employees in this area. Recent oil industry rumors have it that other oil companies are now looking at the Ecuadorian project with interest. Your company must move quickly or lose its chance.

The ACME Oil Company's minimum to negotiate: If the project is going to happen, it must happen soon. You cannot accept a project below a net value of $10 billion per year. Your company will not accept an unstable situation in the rainforest when it comes to the safety of your employees.

- The Ecuadorian Government: [*Only this group reads this paragraph*]
  Your government permitted the ACME Oil Company to conduct exploration drilling in the remote tropical rainforest of your country with the hopes it would find viable oil reserves... and it did. The ACME proposal is a welcome relief to several decades of economic depression and falling currency values on the world market. The exploration came at a cost. The indigenous peoples of the rainforest (with the help of an international environmental group, Green Voice) have complained about the oil exploration and are vehemently opposing any project. Local land owners in and around the area where the drilling would take place have clashed several times with indigenous peoples over the prospect of an oil project. Green Voice and the indigenous peoples have banded together. Green Voice has recently brought international attention to the fact that many endangered and threatened species are native to the drilling area. Even if these groups were not opposed to the project, your government lacks the necessary funds to make the project happen. The World Bank is interested in providing loans but is concerned with the tension and instability. The ACME Oil Company is counting on you to satisfy the World Bank and secure the loans. Your government analysts estimate the project to be worth billions of U.S. dollars per year.

  The Ecuadorian Government's minimum to negotiate: The oil project must happen. To do this, your government must secure funding from the World Bank. You must also reach an agreement with the indigenous peoples that will win their support of the project.

Lab: Soil Salinization

Introduction: Soil salinization is characterized by the accumulation of excess salts—especially on the soil surface. Soil salinization may occur naturally due to climatic conditions or when the water table is near the surface. Saline soils may also be caused by humans. Whether natural or the result of human activity, soil salinization typically involves the pooling of water at the soil surface that evaporates and leaves behind the dissolved salts and minerals. These salts become more and more concentrated over time. The use of flood irrigation particularly contributes to soil salinization and is made worse when compacted soils have poor drainage. Crops weaken or will not grow in such soils. Irrigation run-off is also saline and will impact waterways from which drinking water may be taken and others draw irrigation water.

Overview: This lab simulates the effect of salinity on germination and plant vigor. A series of trays will each contain a plastic plant starter tray with radish seeds planted in soil. The starter trays will be placed inside a larger tray that can hold water. Watering solutions will be poured into the larger tray and the moisture will be absorbed through the bottom of the starter trays. Each tray will be one trial. The trials will range from a 0.25% to a 4.0% saline solution and distilled water will serve as a control. Over 1 to 2 weeks of observations will be recorded and data taken for percent germination and plant vigor. If desired, the watering solutions can be titrated with silver nitrate to determine the total chlorides in milligrams per liter.

Materials:

1. Materials for set-up: Plastic starter tray for plants, larger tray for watering, potting soil, and radish seeds.
2. Watering solutions (trials): Distilled water (control) and salt solutions of 0.25%, 0.50%, 0.75%, 1.00%, 1.5%, 2.0%, 2.5%, 3.0%, and 4.0%.
3. Titration (optional): 50ml buret, 100ml graduated cylinder, 250ml Erlenmeyer flask, 0.0141 N $AgNO_3$ titrant, and K_2CrO_4 indicator solution.
4. Other: 30cm ruler, calculator, forceps, and magnifying glass or stereomicroscope.

Safety: Goggles and an apron should be worn while handling the titration glassware and solutions. Dispose of waste chemicals in the proper manner.

Procedure:

- Student observations and data page: In a lab notebook or on notebook paper, create places for these observations and data:
    - Time and observations for the set-up for each at a 24-hour period interval
    - Time when germination is first noticed
    - Plant vigor after the seedling has emerged
    - Time and observations for the post-lab procedure
    - (*Optional*) Calculated total chlorides for each solution
- Initial set-up: Fill the cells of the starter tray will soil and gently press down on the soil in each cell with your fingers. Press down the soil approximately 1–2cm and leave a small depression in the middle for the seed. Do not press too hard—the soil should not be compacted. Put one radish seed in each depression and cover it with about 1cm of additional soil. Do not press down on the soil covering. Place the starter tray in the center of the larger watering tray. The trays should be in a place that is warm (25–30°C) and has access to a light source. Pour in the watering solution until the water level rises approximately 2cm on the starter tray. If every tray is the same size, this will ensure an equal volume of watering solution per trial. If there are different tray sizes, you will need to adjust watering volumes to make sure each trial receives a comparable amount of water.
- Pre-germination procedure: Check the moisture in trays every 24 hours. The starter trays should not be allowed to sit in perpetual standing water. It is okay if the bottom of the tray becomes dry in a 24-hour period. The addition of more watering solution should be assessed by gently touching the soil. The soil should feel moist—not dry or damp. If more watering solution is required, make sure each trial gets the same volume of water. Record your observations at each 24-hour cycle, noting the appearance of the soil and looking for signs of germination.
- Post-germination procedure: When the first signs of germination are observed, record the day this occurred. Continue to monitor the moisture and watering schedule as before. As the seedling emerges from the soil, record its height above the soil profile in centimeters. This datum will measure plant vigor. Continue to record observations as before, but now record observations of the seedling as well.
- Post-lab procedure: When the experiment has been ended, carefully remove ungerminated seeds and germinated seedlings, rinse away soil, and lay them on a paper towel. Use a magnifying glass or stereomicroscope to observe the seeds and seedlings. Record your observations and make sketches of representative seeds or seedlings for each trial. *Optional*: Titrate the watering solutions for total chlorides in mg/L.

- Titration: Use a graduated cylinder to put 100ml of the watering solution to be tested into a 250ml Erlenmeyer flask. Add 1ml of the K_2CrO_4 indicator solution and swirl to mix. The liquid in the flask should have a yellow color. Titrate with 0.0141 N $AgNO_3$ titrant until the liquid in the flask has a consistent light to medium orange color. Multiply the milliliters of titrant used by 40 to get total chlorides in milligrams per liter.

Lab Analysis and Questions:

1. Create a graph showing plant vigor over time for each trial.
2. Discuss the results of the trials. Which trials seemed to do as well as the control? At what concentration of salinity (chlorides) did the trials show signs of reduced germination and vigor?
3. How do you explain the presence (or possibility) of a few ungerminated seeds in the control and other trials?
4. When the ungerminated seeds were removed and observed, what did you notice? Was the seed coat broken? Did a primary root emerge? Was there a difference among any ungerminated seeds in the control and various salt concentrations?
5. In your observations of the seedlings, were there any noticeable differences among the seedlings of the various trials that you might correlate to soil salinity?
6. Explain the effect of a saline solution on a germinating seed in terms of tonicity.
7. Use your library, textbooks, or the Internet to discover how terrestrial halophytes are able to grow in saline soil conditions that mesophytes could not tolerate.
8. How are improper agricultural practices contributing to the salinization of soils? How can changes to agricultural practices reduce the salinization of soils?
9. What were the calculated total chloride values in mg/L for each trial?

Teacher Notes: Soil Salinization Lab

1. Seeds should germinate within 3–5 days of planting and seedlings should emerge after about 7 days. This lab could be stopped at this point or continued for another week so that plant vigor may be measured. Depending on the seeds purchased, one can realistically expect 80–90% germination rates. Germination in this lab refers to the breaking of the seed coat by the primary root.
2. It is advised to use a watering tray not much bigger than the starter tray to avoid making prodigious amounts of watering solution. Plastic starter trays can be cut with heavy scissors to adjust their size as desired. Larger starter trays need larger watering trays, and thus a lot of watering solution. The larger the starter tray, the more difficult it can be to measure tiny seedlings toward the middle of the tray. In this case, it may be easier to cut the starter tray smaller or plant them in every other hole.
3. Students should be able to see accumulations of salt in the watering tray (if the tray is not white in color), starter tray, and soil surface toward the end of 2 weeks.
4. The saline solutions can be made with standard table salt. Their concentration is a weight to volume ratio. For example, a 4% solution will be 40g of NaCl in 960ml of water.
5. The 0.0141 N $AgNO_3$ titrant is made by dissolving 2.395g of $AgNO_3$ in water and diluting it to one liter. It is advised to use an analytical balance to weigh the $AgNO_3$. The K_2CrO_4 indicator solution is made by dissolving 5g of K_2CrO_4 into100ml of distilled water. The end point of the titration is when the yellow color of the flask contents remains an orange color, resembling slightly reddish orange juice. A dark orange or red color is too far. The control serves as a titration blank. If distilled water is used to make the control, then the end point for the control should be reached with one or two drops of the titrant. If more than a few drops are needed to titrate the control, this value should be subtracted from the total volume of titrant used to titrate all of the saline solutions.

6. This lab can also be conducted by making the watering solutions unknowns. The control (water) and saline solutions can be labeled Solution A, Solution B, and so forth. Students don't know if their solution is the control or the exact concentration if it is a saline solution. This option will require students to titrate their solutions to discover the true concentration.
7. If the trays are in a location with poor airflow, they may not experience much evaporation, and therefore, less salt accumulation. To accentuate the evaporation and salt accumulation, put a fan blowing gently over the trays. This may require, however, that you check the trays for moisture twice a day.

Additional Resources

Cengage Learning Resources for:

Starr/Taggart, *Biology: The Unity and Diversity of Life* 15th Edition

- Instructor's Resource Manual
- Test Bank and ExamView
- Transparency Acetates
- PowerLecture
- Resource Integration Guide
- CengageNOW
- Student Study Guides and Workbooks

ABC Videos:

- Biology in the Headlines 2005, 2006, 2007, 2008, 2009, and 2010 on DVD
- Environmental Science in the Headlines 2005, 2006, and 2007 on DVD
- Human Biology in the Headlines 2006 on DVD
- Genetics in the Headlines 2008 and 2010 on DVD

Additional Websites:

Bozeman Science: AP® Biology Video Essentials
https://www.youtube.com/playlist?list=PLFCE4D99C4124A27A

Access Excellence at the National Health Museum:
http://www.accessexcellence.org/

Advanced Placement Biology at the University of Georgia:
http://apbio.biosci.uga.edu/

Brown University, Evolution Resources:
http://www.millerandlevine.com/km/evol

Cell Biology Animation:
http://www.johnkyrk.com/

College Board:
www.collegeboard.com

Dennis Kunkel Microscopy, Inc.:
http://denniskunkel.com/

Explore Biology:
www.explorebiology.com

The Franklin Institute – The Human Heart:
www.fi.edu/learn/heart/index.html

Howard Huges Medical Institute:
www.hhmi.org

Human Anatomy Online:
www.innerbody.com

Molecular Medicine in Action:
www.wellscenter.iupui.edu/MMIA

The Myelin Project:
http://www.myelin.org

National Association of Biology Teachers:
www.nabt.org

National Geographic:
www.nationalgeographic.com

PBS Evolution:
http://www.pbs.org/wgbh/evolution/index.html

Science Magazine:
http://www.sciencemag.org

University of California Museum of Paleontology:
www.ucmp.berkeley.edu

Virtual Pig Dissection:
http://www.whitman.edu/biology/vpd

AP® is a trademark registered by the College Board, which is not affiliated with, and does not endorse, this product.